Graphing Calculator Enhancement
to Accompany Barnett Ziegler

PRECALCULUS: FUNCTIONS AND GRAPHS

CAROLYN L. MEITLER
Concordia University

McGraw-Hill, Inc.
New York St. Louis San Francisco Auckland Bogotá Caracas
Lisbon London Madrid Mexico Milan
Montreal New Delhi Paris San Juan
Singapore Sydney Tokyo Toronto

This book is printed on recycled, acid-free paper containing a minimum of 50% recycled de-inked fiber.

Graphing Calculator Enhancement
to Accompany Barnett/Ziegler:
PRECALCULUS: FUNCTIONS AND GRAPHS

ISBN 0-07-041367-3

2 3 4 5 6 7 8 9 0 WHT WHT 9 8 7 6 5 4 3

PREFACE

This manual is written to help you use the power of a graphing calculator to learn mathematics. Examples illustrate how to use the calculator with all topics covered in the textbook.

The ***Graphing Calculator Enhancement*** is to accompany *Precalculus: Functions and Graphs*, by R. A. Barnett and M. R. Ziegler. The chapters of the manual correspond with those of the textbook.

Each chapter contains examples similar to problems in the corresponding section of the textbook. Detailed solutions show how a graphing calculator can be used to solve the problem.

Appendices

There are three appendices. Each of the appendices contains keystrokes for a particular calculator.

Appendix A	TI-81 Graphing Calculator Basic Operations
Appendix B	Casio *fx-7700G* Graphing Calculator Basic Operations
Appendix C	TI-85 Graphing Calculator Basic Operations

Examples are included that illustrate all the keystrokes necessary to solve the problems in the textbook. Many examples show more than one method of solution using the calculator. No programming is used.

Sections and examples in the appendices correspond between appendices so that instruction can be given to a class of students in which more than one calculator model is being used. The same example is in all three appendices.

Graphs

Graphs are intended to be representative of the screen display on the calculator. Hence the curves shown in the figures of this manual are not smooth but rather are similar to the graphing calculator display.

All graphs enclosed in rectangles illustrate what is displayed on the calculator screen. Graphs without enclosing rectangles illustrate concepts of the example but may have additional information shown.

Calculations
All calculations in this manual have been performed using the default setting on the calculator. After all calculations have been completed, numbers are rounded to the desired number of significant digits or number of decimal places as the problem requires. Note that the calculators can be set to a given number of significant digits or a given number of decimal places prior to performing the calculations. This might be convenient in many cases; however it was not done in the examples of this manual. Directions on how to set the calculator are given in Appendices A-20, B-20, and C-20 of this manual.

Built-in Calculator Functions
Most solutions to examples in the chapters in this manual do not use the built-in functions (features) of a calculator. Discussion of some of the built-in features can be found in the appendix pertaining to a particular calculator.

Exercise sets
Exercise sets are found at the end of each chapter in the manual. Many exercises refer to textbook problems. Occasionally additional problems are included that use graphing calculator capabilities. The problems in the text that are designated as graphing calculator problems are not included in the exercise sets of this manual.

ACKNOWLEDGEMENTS

The completion of this project would not have occurred without the support and assistance of several individuals. First, and foremost, are the textbook authors, Michael R. Ziegler and Raymond A. Barnett. Their suggestions were most helpful in forming the philosophy, the layout, and the content of this manual. Also I would like to thank Mary Frenn who patiently read the manuscript and worked through all of the problems. Her help in checking for accuracy and consistency was invaluable. Finally, I thank my family for their encouragement and support.

Carolyn L. Meitler

CONTENTS

CHAPTER 1

EQUATIONS AND INEQUALITIES

Example 1 Textbook Section 1-1

Show that $x = \sqrt{3.52}$ is a solution of $x^3 - 3.4x^2 - 1.08x + 9.096 = 2.4x^2 + 2.44x - 11.32$.

Solution: A graphing calculator allows you to store expressions as well as the values for the variables. These stored values and expressions can be easily changed allowing expressions to be evaluated for different values of the variables. This example asks you to evaluate the expressions on both sides of the equal sign to see if they are the same for this value of x. (See Appendix Sections A-5, B-5, or C-5 of this manual. These sections show how to store expressions and values for variables in the calculator and how to evaluate expressions.)

Store the expression on the left side of the equal sign as X^3-3.4X^2-1.08X+9.096. Store the expression on the right side of the equal sign as a second function 2.4X^2+2.44X-11.32. Store the value of X as √3.52 and evaluate the expressions. The displayed value of each of the expressions is 1.705845782. Hence this value of x is a solution to the equation.

Example 2 Textbook Section 1-1

Show that $\frac{x^2-1}{x-1} = x+1$ is not true for all values of x.

Solution: Store 1 as the value of X in the calculator. Enter the left-hand expression as (X^2-1)÷(X-1) and execute. (See Appendix Section A-5, B-5 or C-5 of this manual.) An error will occur since the left-hand expression is not defined for $x = 1$. Hence the expressions are not equal for all values of x. [Note that in this example we did not store the expression before evaluating it. It is not necessary to store an expression before evaluating it. However, it is usually a good idea to store the expression so as to save time when evaluating it for several different values of the variables.

Example 3 Textbook Section 1-1

Approximate the solution to two decimal places to $3x - 2(2x - 5) = 2(x + 3) - 9$.

Solution: A graphing technique for solving this equation is to first set the equation equal to zero. Graph y = (the expression) and find the x intercepts. These values of x is the solution to the equation.

Store the expression in the calculator as a function as 3X-2(2X-5)-2(X+3)+9. Graph using the standard or initial RANGE settings. (See Appendix Section A-7, B-7, or C-7 of this manual.) Use zoom and trace to approximate the x intercept to two decimal places. The x intercept is approximately 4.33.

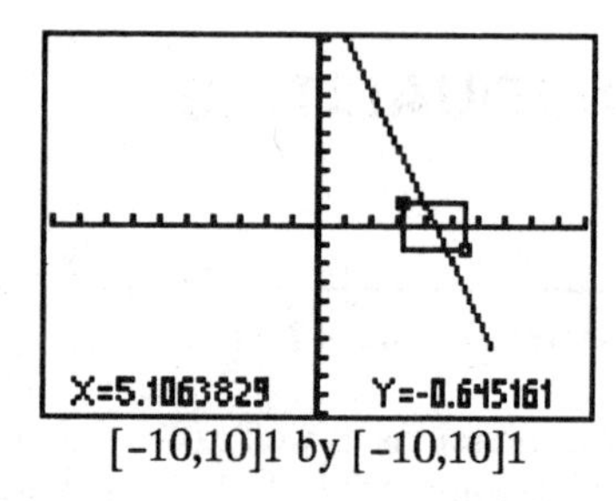

[-10,10]1 by [-10,10]1

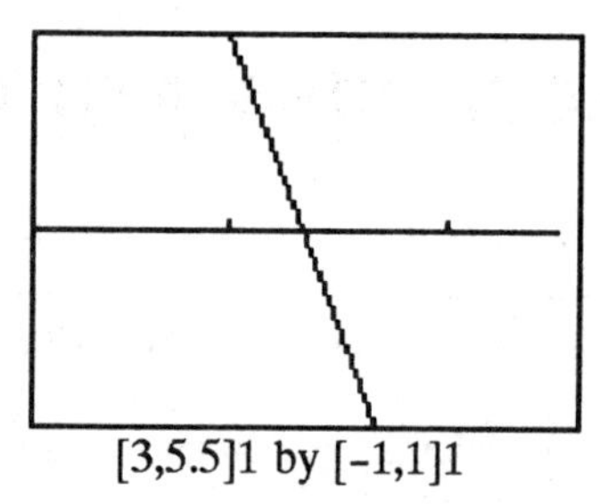
[3,5.5]1 by [-1,1]1

Example 4 Textbook Section 1-1

Solve $\dfrac{5x-22}{x^2-6x+9} - \dfrac{11}{x^2-3x} - \dfrac{5}{x} = 0$ using graphing techniques.

Solution: A graphing technique for solving this equation is to let y equal the left side of the equation and graph the resulting function. We want to find the x values that make the expression (or y) equal to zero. This is the same as finding the x intercepts of the function. (See Appendix Section A-11, B-11, or C-11 of this manual.)

Store the expression in the calculator as (5X−22)÷(X^2−6X+9)−11÷(X^2−3X)−5÷X . Graph using the standard or initial RANGE settings. (See Appendix Section A-7, B-7, or C-7 of this manual.) The figures below illustrate how the features of the graph can be examined more closely by using three different RANGE settings. (See Appendix Section A-8, B-8 or C-8 of this manual.) First use [-5,0]1 by [-.05,.05].01. Now use zoom and trace to find the x intercept of -4. Then set the RANGE at [0,4]1 by [-20,5]2 to observe that there are no x intercepts for $0<x<4$. Finally, set the RANGE at [4,15]1 by [-2,1].5 to observe that the graph gets close to the x axis but never crosses it. Hence there is only one solution to this equation, at $x = -4$.

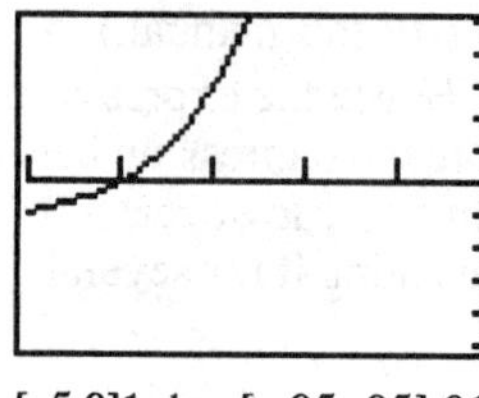
[-5,0]1 by [-.05,.05].01

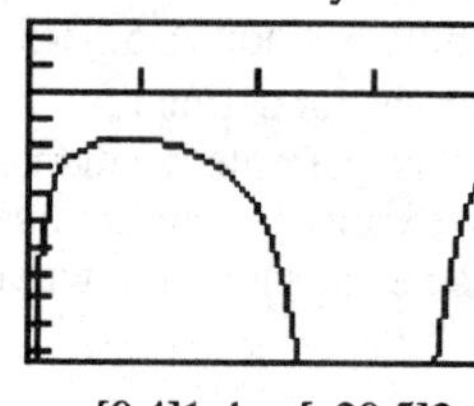
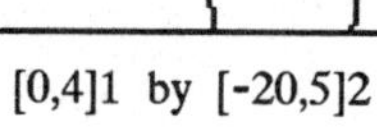
[0,4]1 by [-20,5]2

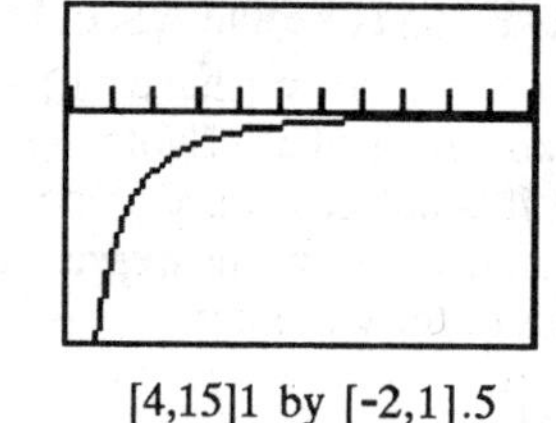
[4,15]1 by [-2,1].5

Example 5 Textbook Section 1-2

A rental company rents a piece of equipment for $60.00 flat fee plus $5.00 per hour.

(A) Make a chart showing the number of hours the equipment is rented and the cost for renting the equipment for 2 hours, 5 hours, 10 hours time, and x hours time.

(B) Write an algebraic expression representing the cost y as a function of the number of hours x. Assume x can be measured to any decimal portion of an hour. (In other words, assuming x is any nonnegative real number.)

(C) Graph the expression from Part (B).
(D) Use the graph to approximate, to one decimal place, the number of hours the equipment was rented if the bill is $216.25 before tax.

Solution:
(A)

x Hours rented	y Cost before tax
2	60 + 5(2) = 70
5	60 + 5(5) = 85
10	60 + 5(10)=110
x	60 + 5 x

(B) Hence $y = 60 + 5x$ where $x \geq 0$.

(C) Note the x and y values in the table help in determining how to set the RANGE values in the calculator. (See Appendix Section A-9, B-9, or C-9 of this manual.) Try [0,24]4 by [0,150]30.

The problem requires that both x and y be nonnegative. The graph is shown at the right.

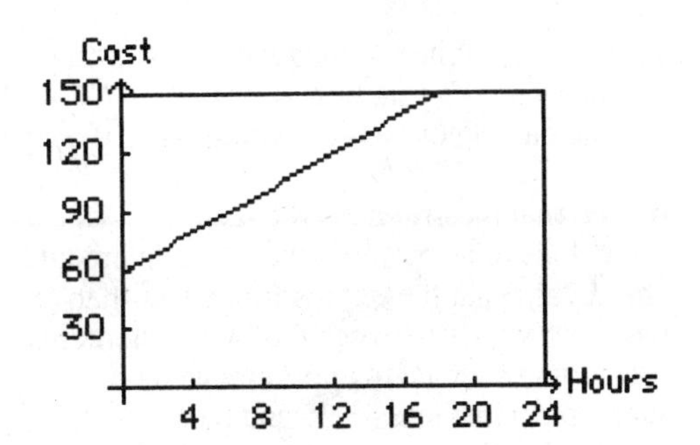

(D) Use the trace and zoom features of the calculator to approximate x (the number of hours rented) when the y (the cost) is 216.25. (See Appendix Sections A-8, B-8, or C-8 of this manual.) The result is 31.25. Hence the equipment was rented for approximately 31.3 hours.

Example 6 Textbook Section 1-1

For what real numbers x does the expression $\sqrt[4]{2x^2-5}$ represent a real number? Express your answer accurate to two decimal places.

Solution: Store the expression in the calculator as (2X^2-5)^(1÷4). Graph using [-5,5]1 by [-5,5]1 as the RANGE settings. (See Appendix Section A-7 & A-8, B-7 & B-8 or C-7 & C-8 of this manual.) We see that the graph is not displayed for all values of x. Use the arrow keys to set the cursor at the end of one piece of the graph where x is near 1.63. Zoom in using this feature of the calculator. (See Appendix Section A-8, B-8, or C-8 of this manual.) Use trace again to see that the y value is now closer to 0 and x is closer to 1.58.

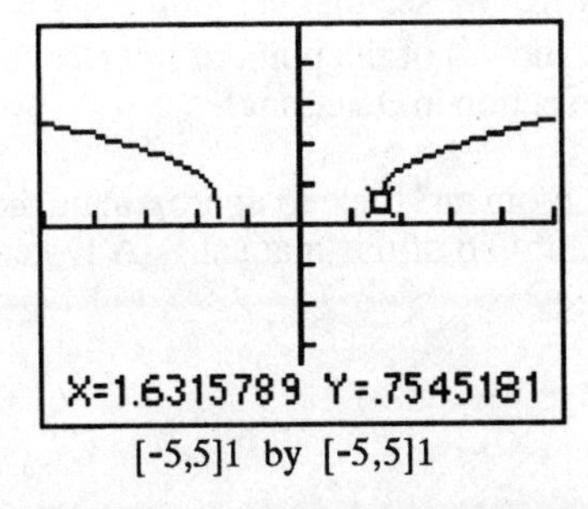

[-5,5]1 by [-5,5]1

Repeated zooming and tracing shows $x \approx 1.58$. Using the symmetry of the graph we can conclude that the function is defined for values of $x < -1.58$ and $x > 1.58$. Algebraically it can be shown that the domain of the function is $\{x \mid 2x^2-5 \geq 0\}$ or $\{x \mid x \geq \sqrt{5/2} \cup x \leq -\sqrt{5/2}\}$.

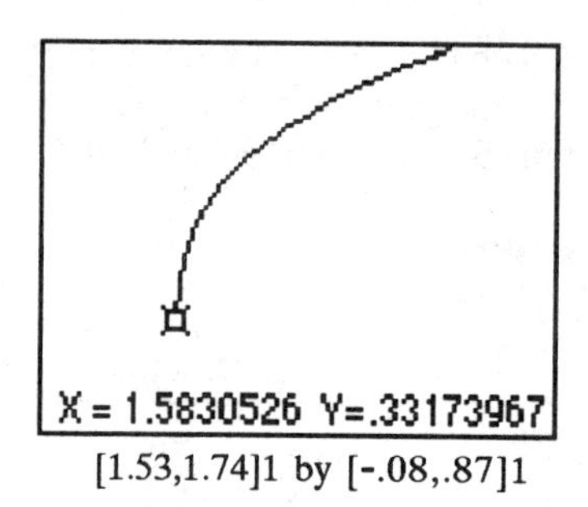

[1.53,1.74]1 by [-.08,.87]1

Example 7 Textbook Section 1-4

Solve $|4.8-2x| < 7.3$. Approximate the solution to two decimal places.

Solution: Write the inequality as $|4.8-2x| -7.3 < 0$. Store the left side of the inequality in the calculator as as a function as ABS (4.8-2X)–7.3 where ABS is the absolute value function built into the calculator. Graph using [-10,10]1 by [-10,10]1 as the RANGE.

We see that the graph of $|4.8-2x| - 7.3$ is below the x axis between the two x intercepts. This means that the expression is less than zero for x values between the two x intercepts. Use zoom and trace to approximate the x intercepts. They are –1.25 and 6.05. Hence the solution to the inequality is (–1.25,6.05) or $-1.25 < x < 6.05$.

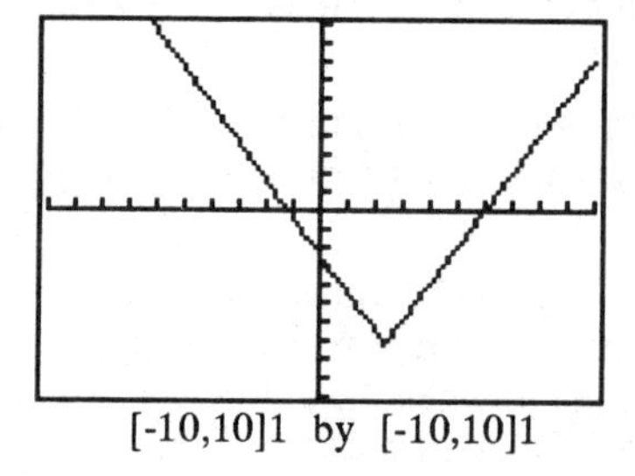
[-10,10]1 by [-10,10]1

Example 8 Textbook Sections 1-3 & 1-4

Solve $|2x^2 + 3x - 2| \geq 4$. Approximate the solution to two decimal places.

Solution: Store the left side of the expression in the calculator as a function as ABS (2X^2+3X–2) where ABS is the absolute value function. Store the right side of the expression as a separate function. (See Appendix Section A-12, B-12 or C-12 of this manual.) Graph $y = |2x^2 + 3x - 2|$ and $y = 4$ on the same coordinate axes using [-5,5]1 by [-5,5]1 as the RANGE. We see that the graph of $y = |2x^2 + 3x - 2|$ is above the graph of $y = 4$ for all values of x to the left of the point of intersection in Quadrant II and to the right of the point of intersection in Quadrant I.

Use zoom and trace to approximate the two points of intersection. (See Appendix Section A-8, B-8, or C-8 of this manual.) A typical zoom box is shown below for the intersection point in

Quadrant I. The approximate coordinates to two decimal places of the points of intersection are (-2.64,4) and (1.14,4). Hence the approximate solution to the inequality is:

$x < -2.64$ or $x > 1.14$ Inequality notation

$(-\infty,-2.64] \cup [1.14,\infty)$ Interval notation

Another method would be to graph $y = |2x^2 + 3x - 2| - 4$ and find the x intercepts of -2.64 and 1.14. The solution to the inequality are the x values where $y \geq 0$.

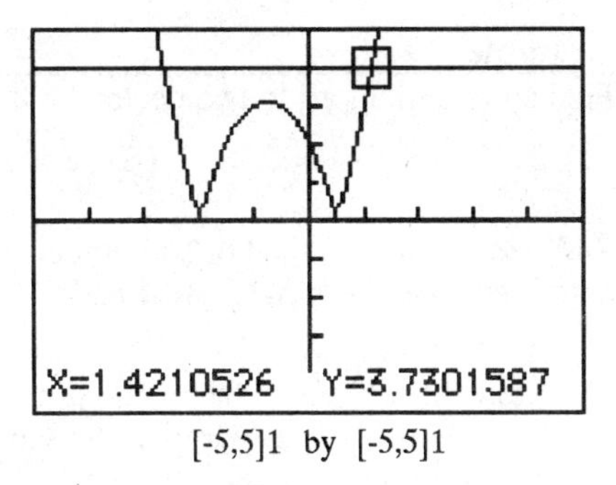

[-5,5]1 by [-5,5]1

Example 9 Textbook Section 1-6

Determine the number of real roots (or zeros) and imaginary roots (or zeros) of each of the following by graphing:

(A) $f(x) = 3x^2 - 2x + 5$ (B) $g(x) = -3x^2 - 7x - 4$

Solution: We wish to find the values of x at which each of these functions cross the x axis since this is where the functions have the value of zero. In other words, we wish to solve $3x^2 - 2x + 5 = 0$ and $-3x^2 - 7x - 4 = 0$. We will graph both parts of this problem at one time. Store the function of Part (A) as 3X^2-2X+5 and the function of Part (B) as -3X^2-7X-4 as separate functions in the calculator. Graph using a RANGE setting of [-10,10]1 by [-10,10]1.

(A) We see that the graph of $y = 3x^2 - 2x + 5$ does not cross the x axis. Therefore there are no real roots. There are two imaginary roots. These can be found using the quadratic formula.

(B) The graph of $y = -3x^2 - 7x - 4$ appears to just touch the x axis. However when this area is enlarged using zoom we see the curve actually crosses the x axis in two places. (See Appendix Section A-8, B-8, or C-8 of this manual.) Hence there are two real roots (or zeros).

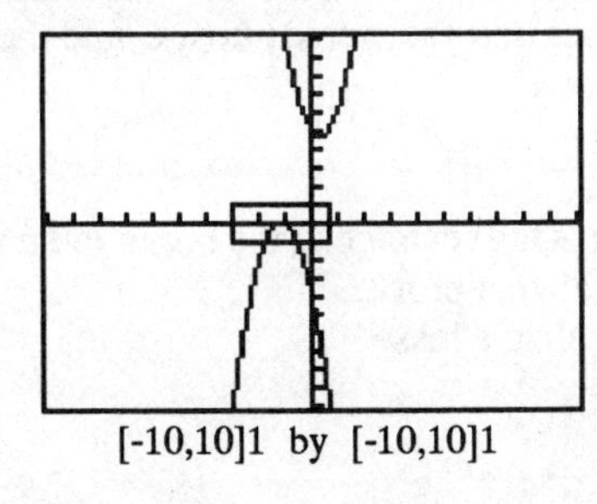

[-10,10]1 by [-10,10]1

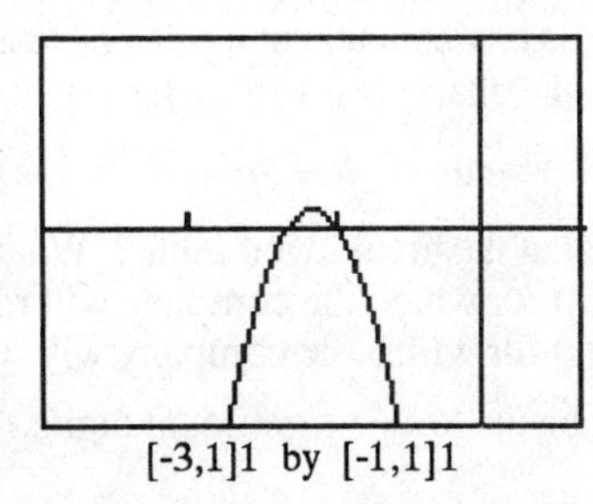

[-3,1]1 by [-1,1]1

Example 10 Textbook Section 1-7

Find all real solutions to two decimal places:

(A) $3x^{-4} + 6x^{-2} - 1 = 0$ (B) $\sqrt{5 - 2x} - \sqrt{x + 6} = \sqrt{5}$

Solution: The method of solution is described in Sections A-11, B-11 and C-11 of this manual.

(A) Store the left side of the expression as 3X^-4+6X^-2-1. Graph. We see two places where the graph crosses the x axis. Hence there are two real roots. Use trace and zoom to find the x intercepts. (See Appendix Section A-8, B-8, or C-8 of this manual.) The solutions to the equation are: $x \approx -2.54$ and $x \approx 2.54$.

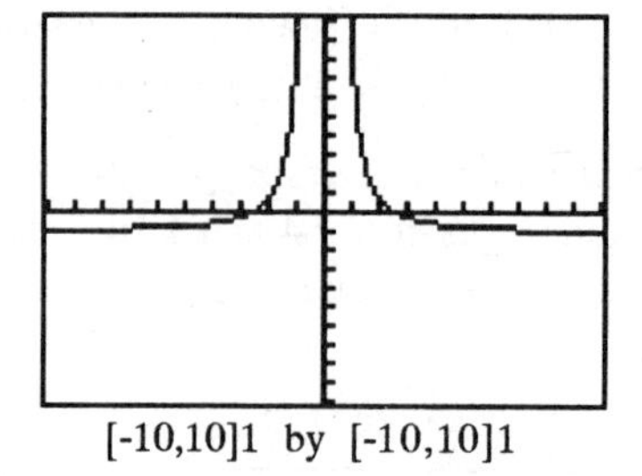

[-10,10]1 by [-10,10]1

(B) Store the left side of the expression as √(5-2X)-√(X+6) and the right side as √5 as separate functions in the calculator. Graph. We see that there is there is only one point of intersection. Use trace and zoom to find the x value of this point. (See Appendix Section A-8, B-8, or C-8 of this manual.) The solution to the equation is $x \approx -4.07$.

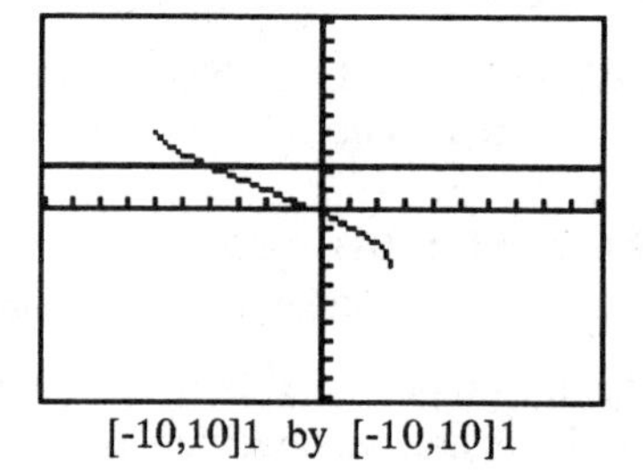

[-10,10]1 by [-10,10]1

Example 11 Textbook Section 1-8

A company manufactures and sells flashlights. For a particular model, the marketing research and financial departments estimate that at a price of $\$p$ per unit the weekly cost C and the revenue R (both in thousands of dollars) is given by the equations

$$C = 7 - p \qquad \text{and} \qquad R = 5p - p^2$$

(A) What is the cost at the break-even point? What is the revenue at the break-even point?
(B) Find the price(s) for which the company will realize a profit.
(C) Find the price(s) for which the company will realize a loss.

Approximate all answers to two significant digits.

Solution: Use x in place of p and graph $y = 7 - x$ and $y = 5x - x^2$ on the same coordinate axes. Store the functions as two separate functions 7-X and 5X-X^2 in the calculator.

(A) Use trace and zoom to find the intersection points (1.6, 5.4) and (4.4, 2.6). (See Appendix Section A-8, B-8, or C-8 of this manual.) The cost and the revenue are equal at these points. These are the break-even points.

A price of \$1.60 will result both in a cost and revenue of \$5,400. A price of \$4.40 will result in both a cost and revenue of \$2,600.

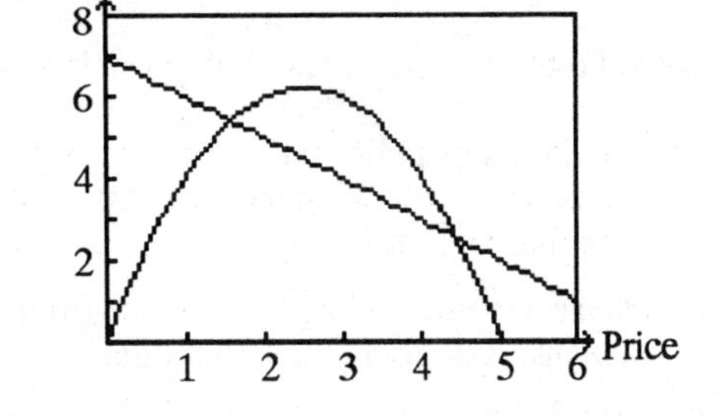

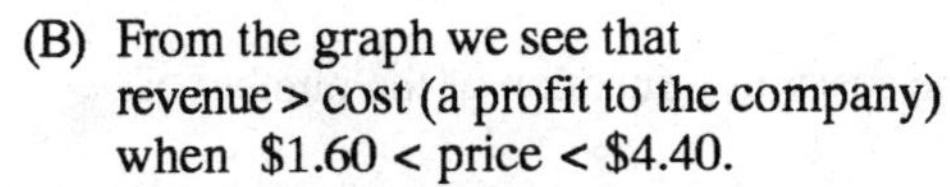

(B) From the graph we see that revenue > cost (a profit to the company) when \$1.60 < price < \$4.40.

(C) From the graph we see that cost > revenue (a loss to the company) when $p<\$1.60$ and $p>\$4.40$. Since p cannot be negative, we must exclude the values $p \le 0$. Thus a loss occurs for $\$0 \le p < \$1.60$ or $p > \$4.40$.

Example 12 — Textbook Section 1-8

Solve to two decimal places $\frac{x^2-3x-10}{1-x} \ge 2$.

Solution: Subtracting 2 from both sides of the inequality gives $\frac{x^2-3x-10}{1-x} - 2 \ge 0$. Store the left side of this inequality as a function in the calculator as (X^2-3X-10)÷(1-X)-2. Graph using [-10,10]1 by [-50,50]5. We see that the graph of the left side of the inequality is above the x axis in Quadrant II; that is, to the left of the x intercept on the negative x axis. Also, it is above the x axis in Quadrant I. Use trace and zoom to find the x intercepts of -3.00 and 4.00. Now use trace to see that the y values change from negative to positive numbers when the x values change from a little less than 1.00 to a little greater than 1. Examining the function we see it is not defined when x=1.00. Hence we conclude that the inequality is true for $x<-3$ or $1<x<4$. This can also be written as $(-\infty,-3.00) \cup (1.00,4.00)$.

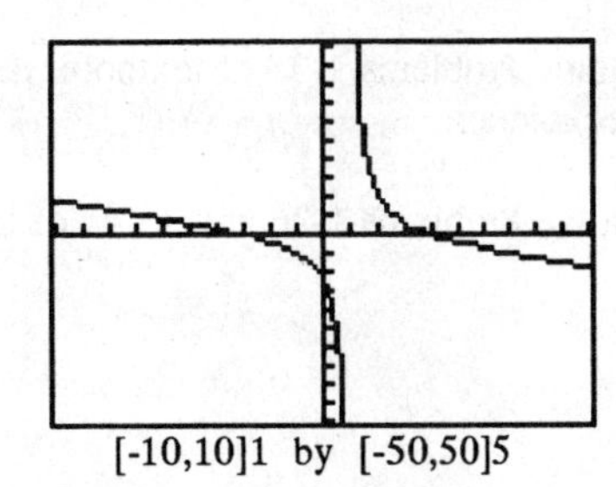

[-10,10]1 by [-50,50]5

▲▲▲ EXERCISE 1

Use two decimal place accuracy when solving the following problems using graphing techniques.

1. Solve Problems 1-47 of textbook Exercise 1-1 using graphing techniques.

2. Refer to Problem 11 of textbook Exercise 1-2.
   (A) Make a table showing the cost of producing 2000, 4000, 6000 and 8000 record albums. Use four significant digits.

   (B) Make a table showing the revenue from selling 2000, 4000, 6000 and 8000 record albums. Use four significant digits.

   (C) Write an algebraic expression representing the cost y as a function of the number of record albums x that are produced.

   (D) Write an algebraic expression representing the revenue y as a function of the number of record albums sold.

   (E) Graph both functions on the same coordinate axes.

   (F) Use trace and zoom to find the break-even point.

   (G) Use the graph to determine how many record albums need to be made to have a revenue of $55,000 ? How much profit is made for this number of record albums?

3. Solve Problems 43-54 of textbook Exercise 1-3. Change the variable to x and use graphing techniques.

4. Solve Problems 39-60 and 83-92 of textbook Exercise 1-4 using graphing techniques.

5. Solve Problems 7-52 of textbook Exercise 1-6 using graphing techniques.

6. Solve Problems 1-38 of textbook Exercise 1-7. Change the variable to x and use graphing techniques.

7. Solve Problems 1-26 and 33-44 of textbook Exercise 1-8 using graphing techniques.

CHAPTER 2

GRAPHS AND FUNCTIONS

Example 1 Textbook Section 2-1

Graph $x^2 + 4y^2 = 36$. Find the value of a to two decimal places if $(a, 2)$ is on the graph.

Solution: Since graphing calculators can only graph functions we solve this equation for y:

$$y = \pm \frac{\sqrt{36-x^2}}{2}$$

Graph y = √(36-X^2)/2 and y = -√(36-X^2)/2 on the same coordinate axes using [-9,9]1 by [-6,6]1. These values were chosen so that the spacing between the scale marks are the same on both axes. This gives better perspective to the graph.

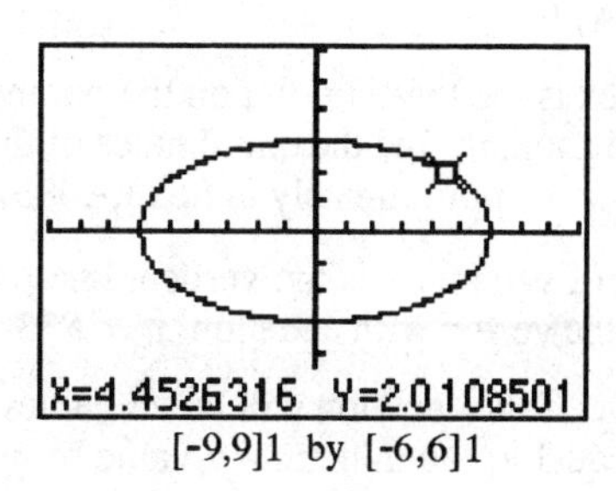

[-9,9]1 by [-6,6]1

Using trace and zoom we find that a = -4.47 or 4.47 correct to two decimal places. (See Example 1 of Appendix Section A-8, B-8, or C-8 of this manual.)

Example 2 Textbook Section 2-2

Find the distance between the points (-52.2, 35.6) and (-90.2,42.8).

Solution: Identify the first point as (A, B) and the second as (C, D). Using these letters the distance formula is $\sqrt{(A-C)^2 + (B-D)^2}$. Store this expression in the calculator as √ ((A-C)^2+(B-D)^2). Store -52.2 as A, 35.6 as B, -90.2 as C and 42.8 as D in the calculator. Evaluate the expression. The distance is 38.68 to two decimal places. (See Example 2 of Appendix Section A-5, B-5 or C-5 and A-13, B-13, or C-13 of this manual.)

Example 3 Textbook Section 2-3

For $f(x) = 8.9x^3 + 1.2x^2 - 5.8\sqrt{x-2} + 19.1$ and x=2.8 and h=.015, find the value of $\frac{f(x+h)-f(x)}{h}$ to two decimal places.

Solution: Store the function as 8.9X^3+1.2X^2-5.8√ (X-2)+19.1. (See Appendix Section A-5, B-5, or C-5 of this manual). Calculate $x+h$ and store as X in the calculator. The result is 2.815. Evaluate the function at this value of X and store as another variable, say B. B≈221.90. Now store 2.8 as X in the calculator and .015 as H. Evaluate the function at this value of X and store as another variable, say C. C≈218.69. Finally calculate the quantity (B-C)÷H. The result is 213.96, accurate to two decimal places.

Example 4 Textbook Sections 2-4

Given $f(x) = 3x^2 - 5x + 2$. Find the:

(A) vertex.
(B) equation of the axis of symmetry.
(C) range of the function.
(D) x intercept.
(E) y intercept.

Approximate all answers to three decimal places.

(A)

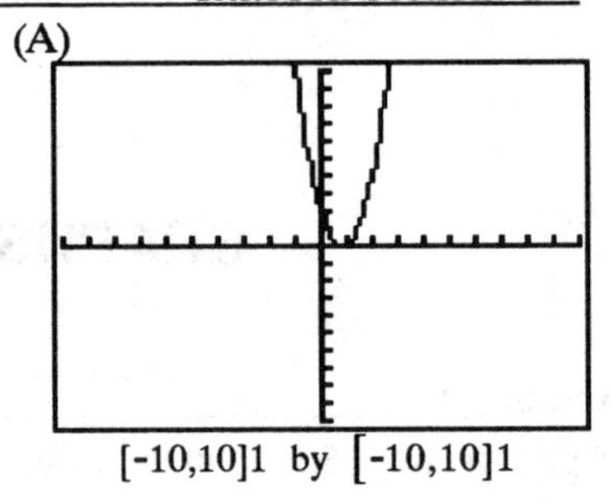

[-10,10]1 by [-10,10]1

Solution: Store the function in the calculator as 3X^2-5X+2. Graph using [-10,10]1 by [-10,10]1. (See Figure (A).)

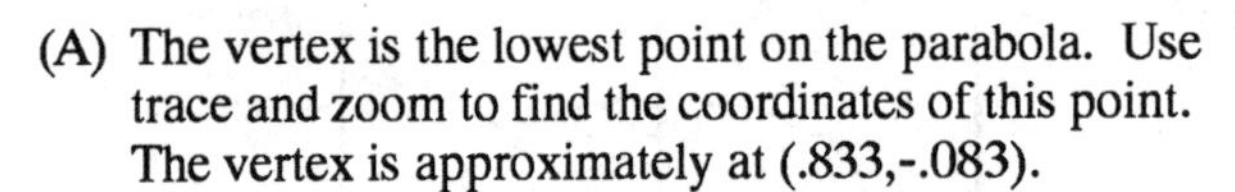

(A) The vertex is the lowest point on the parabola. Use trace and zoom to find the coordinates of this point. The vertex is approximately at (.833,-.083).

(B) The axis of symmetry is the vertical line passing through the vertex with equation $x = .833$.

(C) The range of the function will be all values of y greater than or equal to the minimum y value or $y \geq -.083$.

(B)

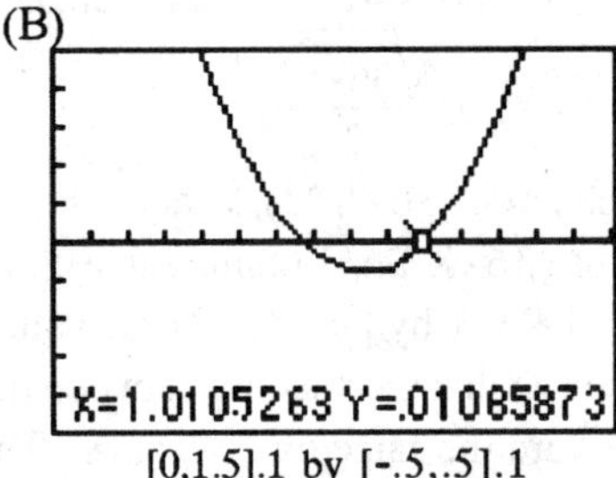

[0,1.5].1 by [-.5,.5].1

(D) The x intercepts can be found by tracing and zooming. (See Figure (B).) They are approximately .667 and 1.000 .

(E) The y intercept can be found be evaluating the function when $x = 0$. Store 0 as X and evaluate the function. The y intercept is 2.000 .

Example 5 Textbook Section 2-4

Graph $y = \begin{cases} x^2 + 1 & x \leq 1 \\ x^2 - 2 & x > 1 \end{cases}$

Find the domain, range and points of discontinuity.

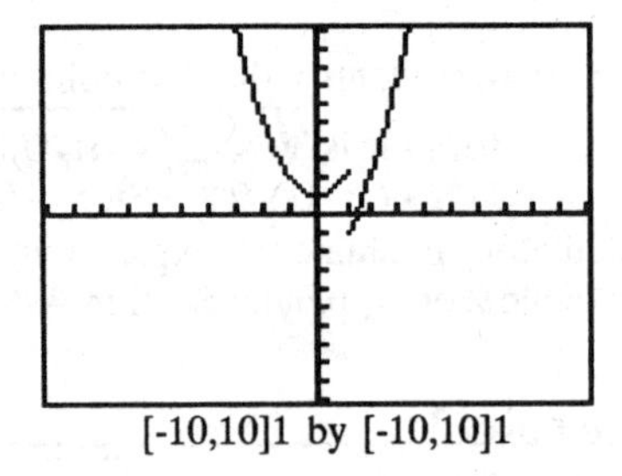

[-10,10]1 by [-10,10]1

Solution: Using Method 2 of Appendix Section A-10, B-10, or C-10 of this manual, we graph $y = x^2 + 1$ for $x \leq 1$ and $y = x^2 - 2$ for $x > 1$. (See figure to the right.) Examining both pieces of this function shows that there is no value of x for which it is not defined. Hence the domain is all real numbers.

Use trace and zoom to see that the minimum value of y is a little greater than -1 but never equal to -1 since $x > 1$ when $y = x^2 - 2$. Hence the range is $(-1, \infty)$.

Since there is a break at $x = 1$ the graph is discontinuous at $x = 1$.

The display on the calculator does not show which endpoint is included. This must be determined from the algebraic expression of the function. The right-hand endpoint of the left-most piece of the graph is included since x is less than <u>or</u> equal to 1. The left-hand endpoint of the rightmost piece of the graph is not included since x is greater than but not equal to 1.

Example 6 Textbook Sections 2-2 & 2-4

Graph $f(x) = x + \frac{|x-3|}{x-3}$.

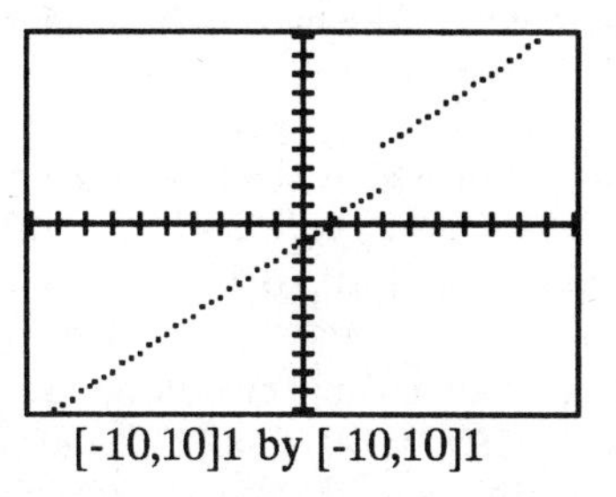
[-10,10]1 by [-10,10]1

Write this function as a piece-wise defined function without using absolute value signs. Find the domain, range and any points of discontinuity.

Solution: Store the function as X+(ABS(X-3))/(X-3) and graph using the dotted mode for graphing. We see that a break occurs in the graph at $x = 3$. There is a point of discontinuity at $x = 3$.

Use trace and zoom to find the two intercepts at (0,-1) and (1,0) for the left portion of the graph (when $x < 3$). Using these points we can find the slope: (0-1)/(-1-0) or 1. Using the point-slope form for the equation of a line we have $y - 0 = (1)(x-1)$ or $y = x - 1$.

Use trace to get two points on the graph for $x > 3$. Use these two points to find the slope of the line. The slope is 1. Now use one of the points and the point-slope form of the line to get $y = x + 1$. Hence we have

$$y = \begin{cases} x - 1 & x < 3 \\ x + 1 & x > 3 \end{cases}$$

We must now examine the graph to see what happens at $x = 3$. Store 3 as X in the calculator and evaluate the stored function. An error results. This indicates that the function cannot be evaluated when $x = 3$. This means that the function is not defined when $x = 3$. Hence, the domain is $(-\infty,3)\cup(3,\infty)$.

Use trace and observe the values of y as the x values increase from 0 to 5. You will notice a jump in y values as x changes from a value a little less than 3 to a value a little greater than 3. The y values jump from a little less than 2 to a little more than 4. Since the endpoints are not included in the graph the range is $(-\infty,2)\cup(4,\infty)$.

Example 7 Textbook Section 2-5

Graph $f(x) = x^3 - 2x^2$, $g(x) = (x-3)^3 - 2(x-3)^2$, and $h(x) = (x+5)^3 - 2(x+5)^2$ on the same coordinate axes. Describe the similarities and differences between the graphs.

Solution: Store the functions in the calculator as three separate functions:

```
X^3-2X^2
(X-3)^3-2(X-3)^2
(X+5)^3-2(X+5)^2
```

Graph using [-10,10]1 by [-10,10]1.

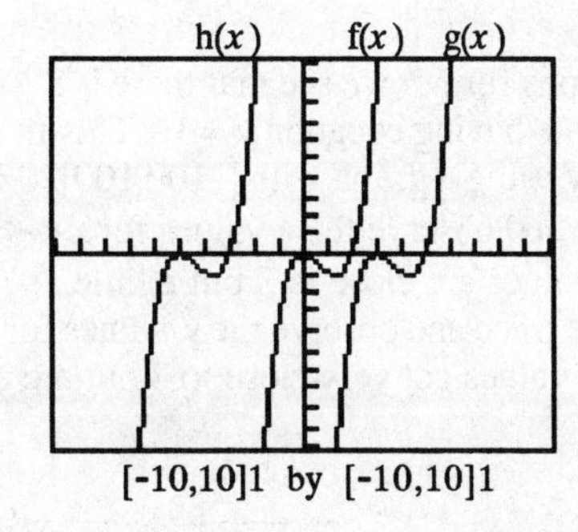

[-10,10]1 by [-10,10]1

The graphs are the same except that $g(x)$ is shifted 3 units to the right of $f(x)$ whereas $h(x)$ is shifted 5 units to the left of $f(x)$.

Example 8 Textbook Section 2-6

Graph $f(x) = \dfrac{3x^2-6x+9}{x^2+x-2}$. Find the x intercepts, y intercepts, vertical asymptotes and horizontal asymptotes. Use two decimal place accuracy.

Solution: Store the function in the calculator as (3X^2-6X+9)÷(X^2+X-2) and graph using [-10,10]1 by [-10,10]1.

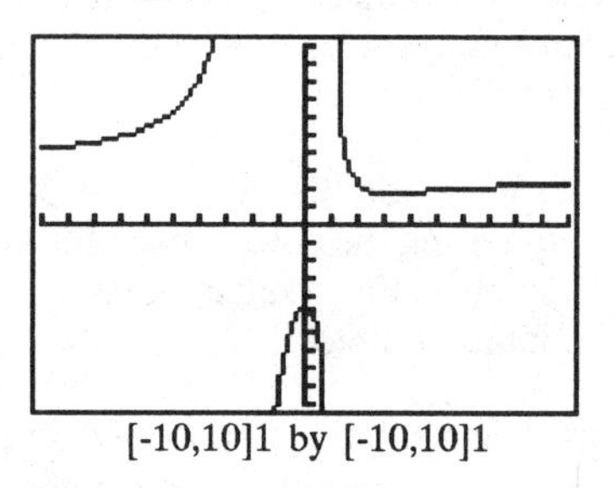
[-10,10]1 by [-10,10]1

There does not appear to be any x intercepts. Set the ZOOM FACTORS to 2 and zoom out once to check this. (See Example 1 Method 2 of Appendix Section A-8, B-8, or C-8 of this manual.)

Examining the graph we see that there is a y intercept. Store 0 as X and evaluate the function to find the y intercept. The y intercept is (0, -4.50).

We also see that there is a vertical asymptote at $x = -2$. This may be more evident if we graph the second and third quadrants alone with a wider view of y values. See the figure to the right using [-10,0]1 by [-50,50]2. The vertical asymptote can be examined further by using trace. Note the values of y jump from a very large positive number to a negative number with large magnitude as x moves from a little to the left of $x = -2$ to a little to the right of $x = -2$.

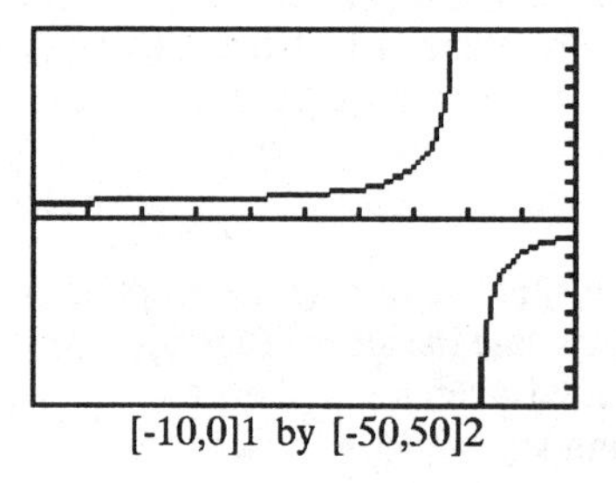
[-10,0]1 by [-50,50]2

Now examine the first and fourth quadrants. We see there is a vertical asymptote at $x = 1$. See the figure to the right using [0,10]1 by [-30,30]5. This vertical asymptote can be examined further by using trace. Note the values of y jump from a negative number with large magnitude to a very large positive number as x moves from a little to the left of $x = 1$ to a little to the right of $x = 1$.

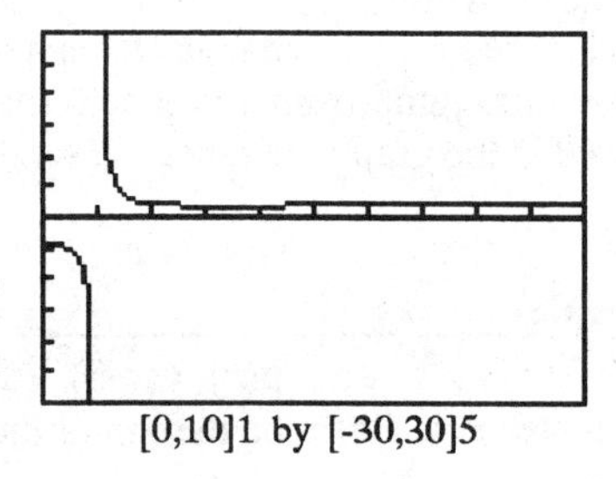
[0,10]1 by [-30,30]5

Examining the graphs further we see that there is a horizontal asymptote having equation $y = 3$. This may be more evident by using a graph with [-100,100]20 by [-5,5]1 . Use trace and observe the y values for $x < -80$. In this case the y values are close to 3 but a little greater than 3. Use trace and observe the y values for $x > 80$. Here the y values get very close to 3 but are a little less than 3.

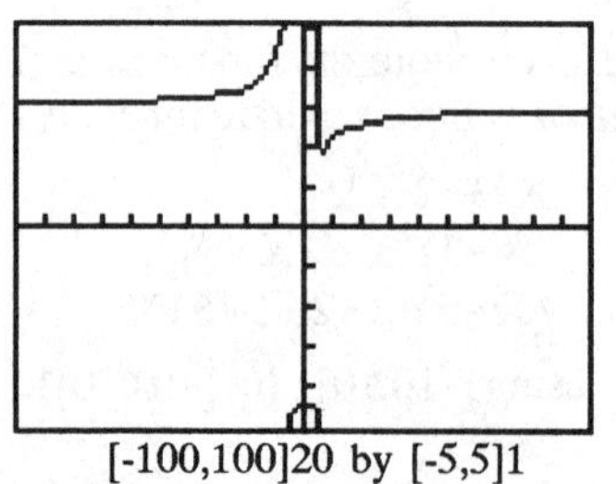
[-100,100]20 by [-5,5]1

Example 9 Textbook Section 2-7

For $f(x) = \sqrt{x+3}$ and $g(x) = 5x^2+4$ find: (A) $(f+g)(2)$ (B) $\left(\frac{f}{g}\right)(-5)$ (C) $(f\circ g)(-1)$

Use two decimal place accuracy.

Solution: Store $y = \sqrt{x+3}$ and $y = 5x^2+4$ as two separate functions in the calculator as √(X+3) and 5X^2+4.

(A) Store 2 as X. Calculate $(f+g)(2)$ by recalling the functions one at a time, adding them algebraically in the calculator and executing. (See Example 2 Appendix Section A-5, B-5, or C-5 of this manual.) The result is 26.24.

(B) Store -5 as X. Calculate $\left(\frac{f}{g}\right)(-5)$ by recalling the functions one at a time, dividing them algebraically in the calculator and executing. An error will occur since $f(-5)$ is not defined.

(C) Store -1 as X. Evaluate $g(x)$. The result is 9. This is stored in the temporary memory ANS. Store ANS as X in the calculator. Now recall $f(x)$ and evaluate at this new value of X. The result is 3.46.

Example 10 Textbook Section 2-8

Show that $y = 2x^3 - .8x^2 + 4$ is not a one-to-one function.

Solution: Store this function in the calculator as 2X^3 - .8X^2+4. Graph using [-10,10]1 by [-10,10]1.

Use the zoom box feature of the calculator to produce the graph in figure (B) at the right. A typical zoom box is illustrated. See Figure (A) at the right.

When using the zoom box feature of the calculator, a wide short box will display the graph with steeper peaks and valleys so that maximum and minimum points can be found more easily.

The enlarged view shows that this graph does not pass the horizontal line test. Hence, the function is not one-to-one.

(A)

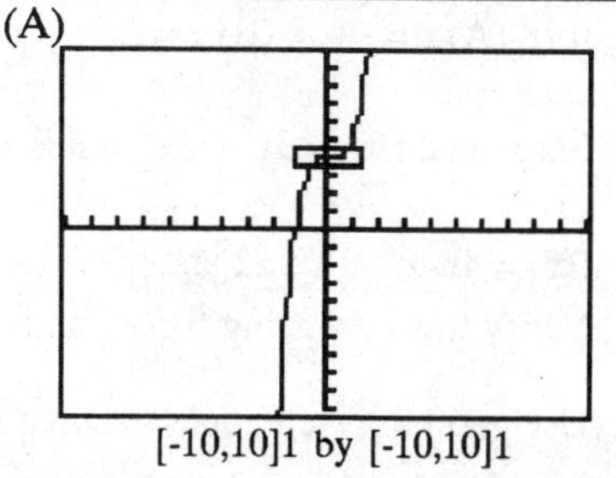

[-10,10]1 by [-10,10]1

(B)

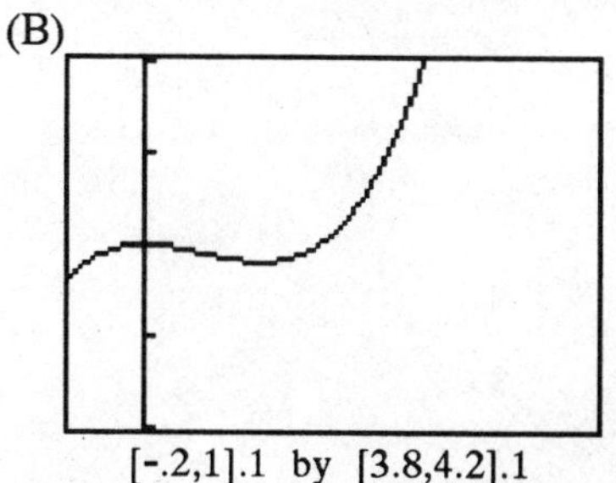

[-.2,1].1 by [3.8,4.2].1

Example 11 Textbook Section 2-8

Find the interval on which the function $y = 2x^3 - .8x^2 + 4$ is decreasing. Use two decimal place accuracy.

Solution: The graph of this function is shown in Example 10. Examining the graph in Figure (B) above we see that the function is decreasing for some values of x. Use trace and zoom to find the maximum point (0, 4) and the minimum point (.27, 3.98). Hence the function is decreasing on the interval (0, .27).

▲▲▲ EXERCISE 2

1. Graph $x^2 + 3y^2 = 12$. Find the value of a to two decimal places if $(a, 1.3)$ is on the graph.

2. Graph $4x^2 + y^2 = 16$. Find the value of a to two decimal places if $(a, -\sqrt{2})$ is on the graph.

3. Graph Problems 7-16, and 63-68 of textbook Exercise 2-2.

4. Use graphing techniques to solve Problems 19-50 of textbook Exercise 2-4.

5. Use graphing techniques to solve Problems 1-10, 23-28, and 39-48 of textbook Exercise 2-5.

6. Find the domain, x intercepts, y intercepts, and all vertical asymptotes, horizontal asymptotes and oblique asymptotes for Problems 1-64 of textbook Exercise 2-6.

7. Evaluate to two decimal places $(f+g)(x)$, $(f-g)(x)$, $(fg)(x)$, $\left(\frac{f}{g}\right)(x)$, $(f \circ g)(x)$, and $(g \circ f)(x)$ for the functions f and g in Problems 41-42 and 51-52 of textbook Exercise 2-7

   for: (A) $x = -1$ (B) $x = 2.5$ (C) $x = .035$

8. Show that $f(x) = 3x^3 - 2x^2 + 3$ is not one-to-one.

9. $f(x) = 4x - x^2$ for $x \geq 2$ and $g(x) = 2 + \sqrt{4-x}$ for $x \leq 4$ are inverse functions of each other. Verify this by finding $f(g(x))$ and $g(f(x))$ when $x = 3.58$.

10. Use graphing techniques to solve Problems 17-22 and 27-32 of textbook Exercise 2-8.

CHAPTER 3

POLYNOMIAL FUNCTIONS: GRAPHS AND ZEROS

Example 1 — Textbook Section 3-1 & 3-2

Evaluate both sides of the equation $\frac{P(x)}{x-r} = Q(x) + \frac{R}{x-r}$ for $P(x) = x^2 + 3x - 7$, $r = 2$, $Q(x) = x+5$, and $R = 3$ at $x = 1$, $x = -3$, and $x = \sqrt{3}$.

Solution: Store (X^2+3X−7)÷(X−2) and (X+5)+3÷(X−2) as separate expressions in the calculator. Store 1 as X and evaluate both expressions. The results should be identical. Repeat for the other values of x. The results should be identical.

Example 2 — Textbook Section 3-2

Fill in the following table by evaluating $P(x) = x^3 + 3x^2 - x - 3$ at the given values of x.

x	P(x)
-4.0	______
-3.5	______
-3.0	______
-2.0	______
-1.0	______
1.0	______
1.5	______
2.0	______

Solution:

Store the expression as X^3+3X^2−X−3

Store the value of x and evaluate the expression.

Repeat for each value of x.

The solutions are:

P(−4.0) = −15 P(−3.5) = −5.625 P(−3.0) = 0

P(−2.0) = 3 P(−1.0) = 0 P(1.0) = 0

P(1.5) = 5.625 P(2.0) = 15

Example 3 — Textbook Section 3-3 & 3-4

Find the zeros of

$P(x) = 2x^5 - 2x^4 + x^2 - \sqrt{2}\,x + .65$

Approximate to two decimal places.

Solution:

Graph $y = 2x^5 - 2x^4 + x^2 - \sqrt{2}\,x + .65$ by storing 2X^5−2X^4+X^2−(√2)X+.65 in the calculator. The parentheses around √2 are needed. Without them the calculator interprets √2X as $\sqrt{2x}$. Graph using [−10,10]1 by [−10,10]1.

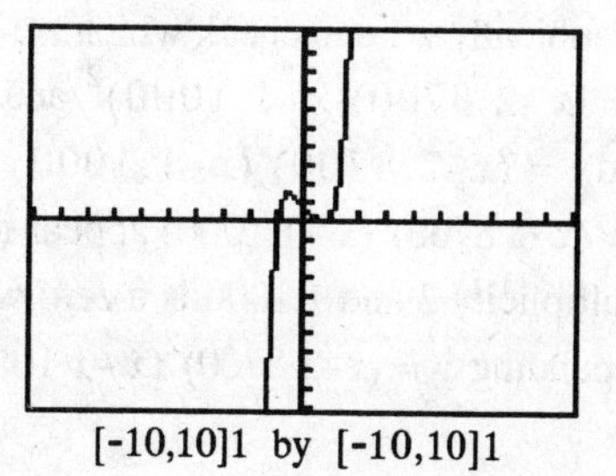

[-10,10]1 by [-10,10]1

By examining the graph we see that the curve crosses the negative x axis. Use zoom and trace to find this zero. The result is $x \approx -.92$.

It is not certain by just examining the graph whether the curve crosses or just touches the positive x axis when using [-10,10]1 by [-10,10]1 as the RANGE setting. Performing several zooms (or setting the RANGE at [.66,.90].01 by [-.02,.02].01) will show there are two more zeros, $x \approx .72$ and $x \approx .82$.

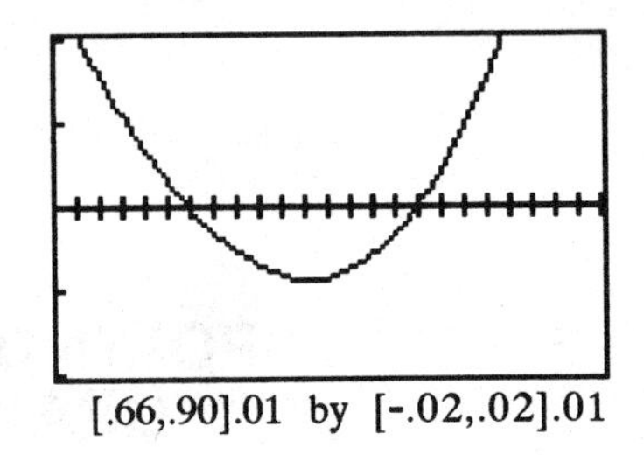

[.66,.90].01 by [-.02,.02].01

Zeros can be checked by storing the value as X and evaluating the expression. For example, store -.92 as X in the calculator and evaluate the function. The result is approximately -.05. The value of y is not exactly equal to zero because we rounded the value of x to two decimal places.

Example 4

Textbook Section 3-4

Find the zeros and their multiplicity of $P(x) = x^3 - 6.84000x^2 + 14.55090x - 9.06059$. Use four decimal place accuracy. Write P(x) in factored form.

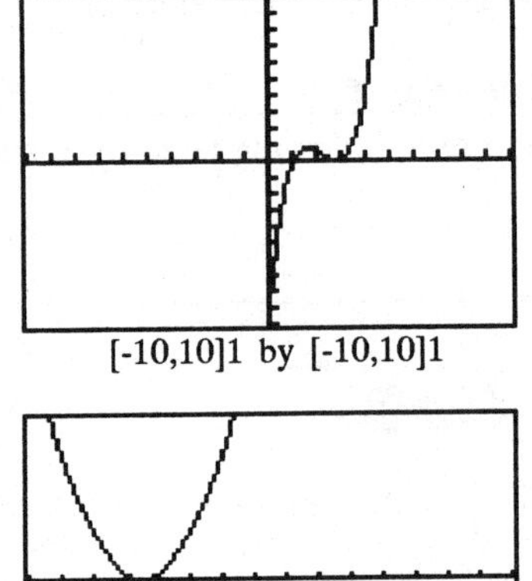

[-10,10]1 by [-10,10]1

Solution: Store the expression as X^3-6.84X^2+14.5509X-9.06059. Use trace and zoom to find the leftmost zero. It is $x \approx 1.1000$.

Set the RANGE at [2.8,3.1].02 by [-.005,.005]1 to see that the curve just touches the x axis between 2.86 and 2.88. Use trace and zoom to find this zero. It is $x \approx 2.8700$. P(x) is a polynomial of degree three which must have either 1 or 3 real zeros. Since we already have two zeros then there must be a total of three real zeros. One of the zeros found in this example must have multiplicity 2.

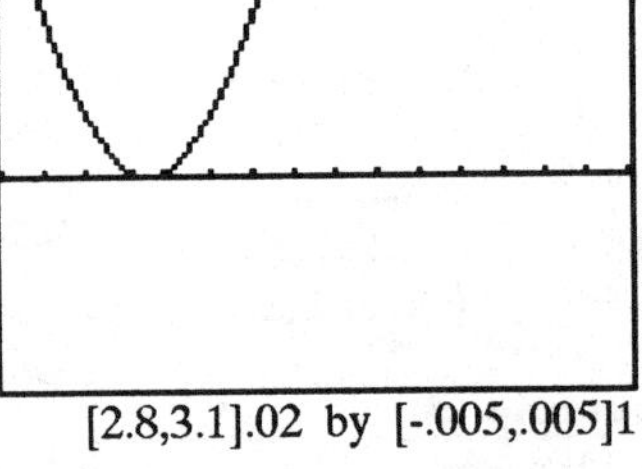

[2.8,3.1].02 by [-.005,.005]1

Graphically we can check which zero has multiplicity 2 by graphing $y = (x-2.8700)(x-1.1000)^2$ and y=P(x) and then comparing the graphs. Then graph y=P(x) and $y = (x-2.8700)^2(x-1.1000)$ and compare the graphs. The graphs of y=P(x) and $y = (x-2.8700)^2(x-1.1000)$ appear identical. Hence the conclusion is that 2.8700 is a zero with multiplicity 2 and 1.1000 is a zero with multiplicity 1. This can be checked algebraically by expanding $y = (x-2.8700)^2(x-1.1000)$.

Example 5 Textbook Section 3-5

Solve $\dfrac{x^2 - 3x - 10}{x^3 - 4x^2 + x + 6} \le 0$.

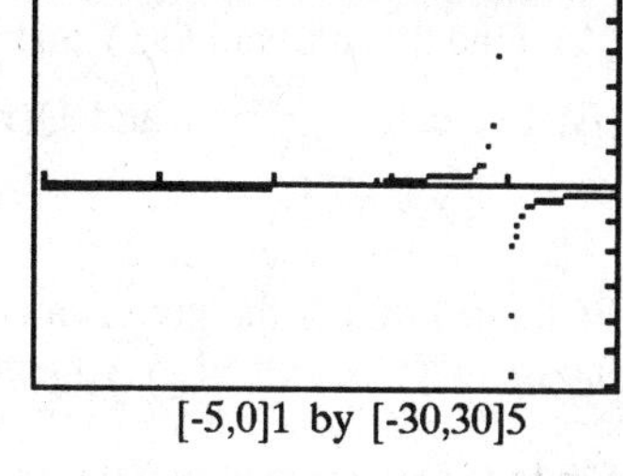

[-5,0]1 by [-30,30]5

Solution: Store the expression as (X^2-3X-10)÷(X^3-4X^2+X+6) and graph using the dotted mode. A first graph using the standard or initial range values will convince you that we need to look at portions of the graph separately. Graph using [-5,0]1 by [-30,30]5. Then graph using [0,5]1 by [-30,30]5.

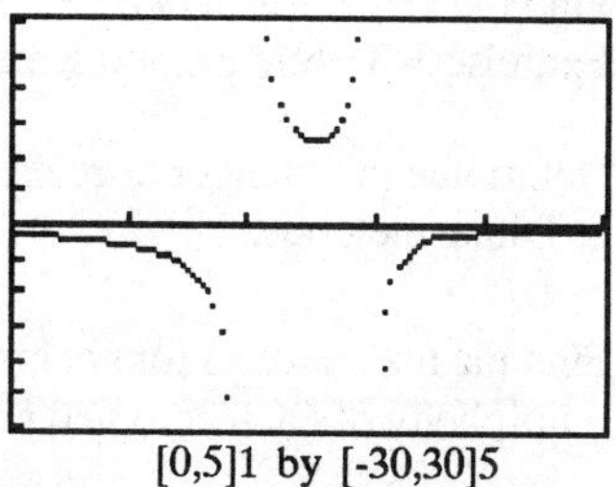

[0,5]1 by [-30,30]5

The solution of the inequality is the set of x values where the graph is on or below the x axis. (See Appendix Section A-12, B-12 or C-12 of this manual.)

For $x < 0$ [-5,0]1 by [-30,30]5
Use trace and watch the y values as you trace when x values are close to -1. The y values jump from a large positive number to a negative number with large magnitude as x moves from the left of -1 to the right of -1. Hence we can conclude that the algebraic expression has value less than 0 for $x > -1$.

We also observe a zero at $x = -2$. Use trace and zoom to verify this. We conclude that the algebraic expression has value less than or equal to zero when $x \le -2$.

For $x > 0$ [0,5]1 by [-30,30]5
Use trace and watch the y values when x values are close to 2. The y values jump from a negative number with large magnitude to a large positive number as x moves from the left of 2 to the right of 2.

Similarly the y values jump from a very large positive number to a negative number of large magnitude as x moves from the left of 3 to the right of 3. Hence we can conclude then that the algebraic expression has value less than 0 for $x < 2$ and $x > 3$.

The solution set of the inequality is $\{x \mid x \le -2, \ -1 < x < 2, \ x > 3\}$.

▲▲▲ EXERCISE 3

1. For Problems 1-36 of textbook Exercise 3-1:
   (A) Find the quotient $Q(x)$ and the remainder R using methods described in the textbook.
   (B) Evaluate $\frac{P(x)}{\text{divisor}}$ and $Q(x) + \frac{R}{\text{divisor}}$ for each problem using $x = -1.2$, $x = 5.3$, and $x = \sqrt{43}$.

2. Evaluate $P(x)$ for the given value of x in Problems 15-18 of textbook Exercise 3-2 using the methods described in Appendix Section A-5, B-5 or C-5 of this manual.

3. Find the real zeros to two decimal places of $P(x)$ in Problems 25-32 of textbook Exercise 3-3 using graphing methods.

4. Determine the number of real zeros of Problems 19-30 of textbook Exercise 3-4 using graphing methods.

5. Find the real zeros to four decimal places of $P(x) = x^4 - 2.5x^3 + 1.5x^2 - .5x - .5$. State the multiplicity of each zero and write $P(x)$ in factored form.

6. Solve Problems 43-46 of textbook Exercise 3-5 using graphing methods.

CHAPTER 4

EXPONENTIAL AND LOGARITHMIC FUNCTIONS

Example 1 Textbook Section 4-1

Complete the table for $f(x) = \frac{1}{2}\left(\frac{1}{4}\right)^x$.

Approximate all values to three decimal places.

x	$f(x)$
-2.2	______
-1.5	______
0	______
1.5	______
2.3	______

Solution:

Store the expression in the calculator as

(1÷2)(1÷4)^X.

(See Appendix Section A-17, B-17 or C-17 of this manual.) Store a value of x and evaluate the expression. Repeat for all values of x.

The solution is: f(-2.2) = 10.556 f(-1.5) = 4.000 f(0) = .500
f(1.5) = .063 f(2.3) = .021

Example 2 Textbook Section 4-1

Graph $y = 2^x$, $y = 3^x$, $y = 5^x$ on the same coordinate axes. Use [-3,3]1 by [-3,3]1. How does a change in base affect the graph of the function?

Solution: Store the functions separately as 2^X, 3^X, and 5^X in the calculator. (See Appendix Section A-17, B-17 or C-17 of this manual.) Graph using [-3,3]1 by [-3,3]1. The graph shows that the function with the larger base is higher (and steeper) for positive values of x and is lower for negative values of x.

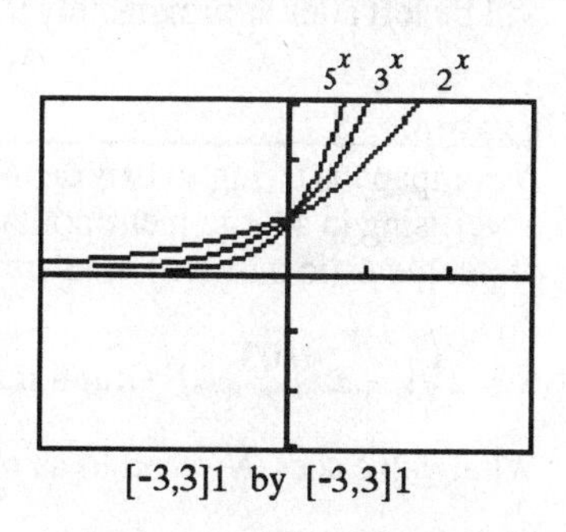

[-3,3]1 by [-3,3]1

Example 3 Textbook Section 4-1

Graph $y = 2^x$ and $y = \left(\frac{1}{2}\right)^x$ on the same coordinate axes.

Use [-3,3]1 by [-3,3]1. What is the relationship between these graphs?

Solution: Store the functions separately as 2^X and (1÷2)^X in the calculator. (See Appendix Section A-17, B-17 or C-17 of this manual.) Set the RANGE and graph. One graph is the reflection of the other about the y axis.

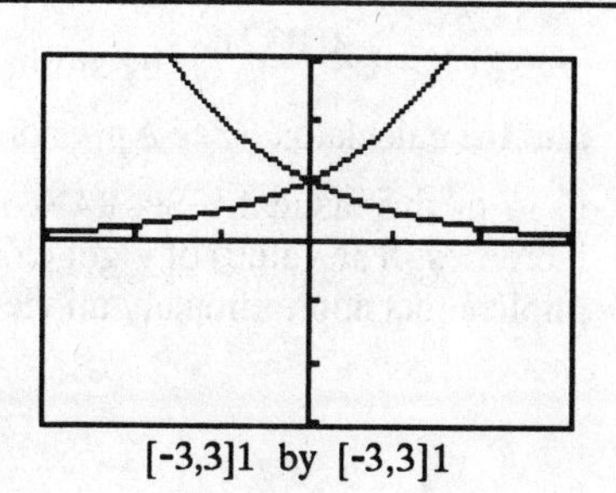

[-3,3]1 by [-3,3]1

Example 4 — Textbook Section 4-1

The radioactive isotope gallium-67 (^{67}Ga), used in the diagnosis of malignant tumors, has a biological half-life of 46.5 hours. If we start with 100 milligrams of the isotope, how many milligrams will be left after 24 hours? When will there be only 25 milligrams left? Approximate answers to one decimal place.

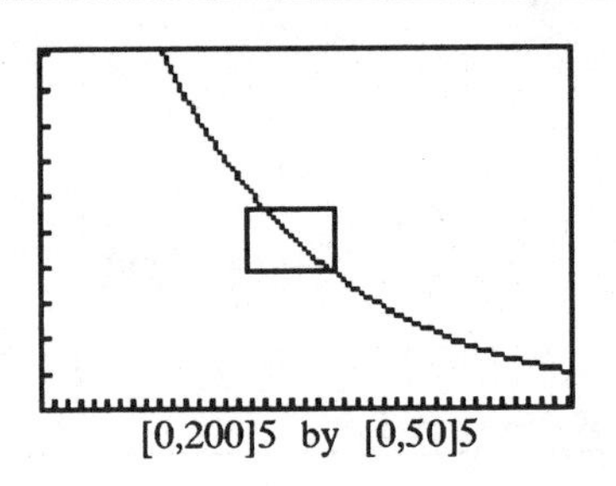

[0,200]5 by [0,50]5

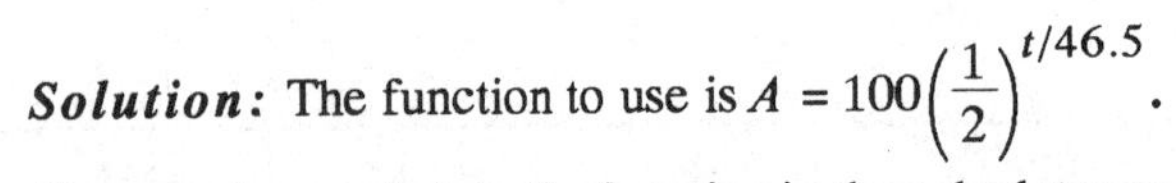

Solution: The function to use is $A = 100\left(\frac{1}{2}\right)^{t/46.5}$.

Change t to x and store the function in the calculator as 100(1÷2)^(X÷46.5). (See Appendix Section A-17, B-17 or C-17 of this manual.)

To find the amount of the isotope remaining after 24 hours, store 24 as X and evaluate the expression. The result is 69.9. Hence, approximately 69.9 mg of the isotope remains after 24 hours.

We need to graph the function in order to approximate when there will be only 25 mg left. Some experimentation may need to be done to set the RANGE. One starting place would be [0,24]1 by [0,70]10 since (24,69.9) is a point on the graph as found above. However we do not see a curve when graphing using this RANGE. We could evaluate the expression for some other values for x and the readjust the RANGE or we could zoom out. Set the RANGE to [0,200]5 by [0,50]5 to get the graph shown above. Now use trace and zoom to find the value of x when $y = 25$. A typical zoom box is shown in the figure. The result is $x = 93.0$. Hence, 25 milligrams of the isotope will be left after approximately 93 hours.

Example 5 — Textbook Section 4-2

A company is trying to expose as many people as possible to a new product through television advertising in a large metropolitan area with 2 million possible viewers. A model for the number of people N (in millions) who are aware of the product after t days of advertising was found to be

$N = 2\left(1 - e^{-0.037t}\right)$ Graph this function for $0 \le t \le 200$.

What value does N tend to as t increases without bound?

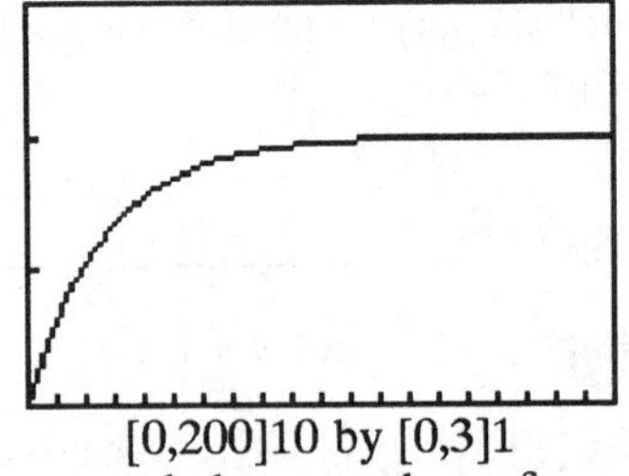

[0,200]10 by [0,3]1

Solution: Change t to x and graph

$y = 2\left(1 - e^{-0.037x}\right)$ by storing 2(1-e^(-.037X)) into the calculator. (See Appendix Section A-17, B-17 or C-17 of this manual.) Set RANGE to [0,200]10 by [0,3]1. Use trace and observe values of y as x increases. The values of y get close to 2. Hence N tends to 2 as t increases without bound. This implies that approximately all viewers will be exposed to this new product in the long run.

Example 6 Textbook Section 4-2 & 4-3

Calculate x to four significant digits in each of the following:

(A) $x = \log\left(3.42 \times 10^{-8}\right)$ (B) $3500 = 2410\left(e^{0.12x}\right)$

Solution: (A) Enter the expression into the calculator directly as [LOG] (3.42×10^[(-)] 8). The scientific mode or engineering mode could also be used for this problem. The answer is -7.466. (See Appendix Section A-20, B-20, or C-20 of this manual.)

(B) Graph $y = 3500$ and $y = 2410\left(e^{0.12x}\right)$ as separate functions. Store the second function as 2410(e^(.12X)). Graph both functions on the same coordinate axes. Adjust the RANGE until you can observe the intersection point. Use trace and zoom to find the x value of the point of intersection. A typical zoom box is shown in the figure. The result is 3.109. Hence the solution to this equation is approximately 3.109.

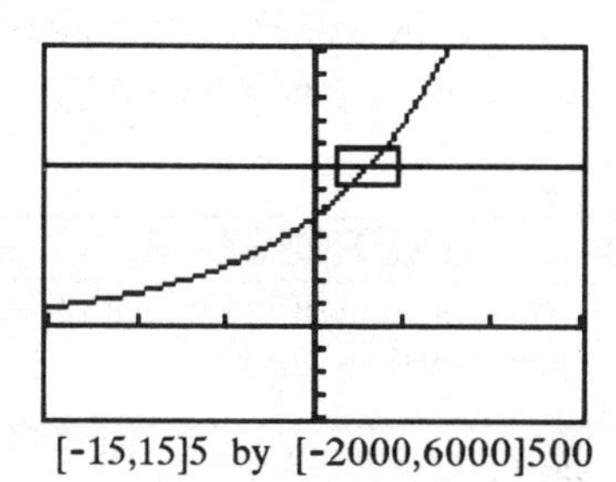

[-15,15]5 by [-2000,6000]500

Example 7 Textbook Section 4-3

Graph $y = 10^x$, $y = \log x$ and $y = x$ on the same coordinate axes. How are the first two graphs related?

Solution: Store the functions as three separate functions in the calculator as 10^X, [LOG] X, and X. Adjust the RANGE and graph. The graph is shown to the right. 10^x and $\log x$ are inverse functions. The line $y = x$ is the line of symmetry.

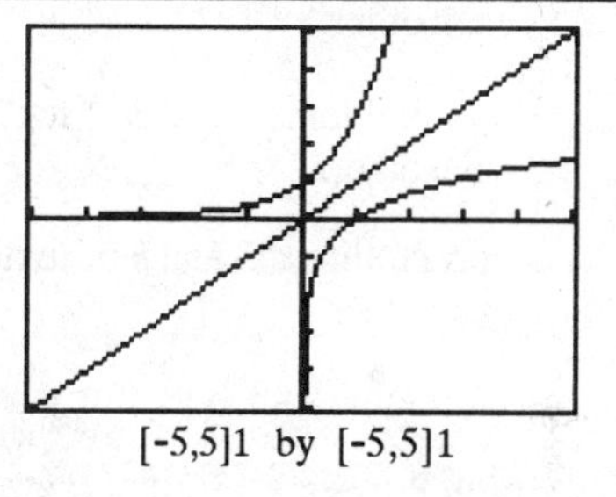

[-5,5]1 by [-5,5]1

Example 8 Textbook Section 4-4

The 1906 San Francisco earthquake released approximately 5.96×10^{16} joules of energy. The magnitude formula is $M = \frac{2}{3} \log \frac{E}{E_o}$ where E is the energy released by the earthquake as measured in joules. $E_o = 10^{4.40}$ joules is the energy released by a very small reference earthquake which has been standardized to this number. What was the magnitude of the 1906 San Francisco earthquake on the Richter scale? Approximate to two decimal places.

Solution: We need to calculate $\frac{2}{3} \log \left(\frac{5.96 \times 10^{16}}{10^{4.40}}\right)$. The calculation can be done directly with no need to store the expression. Use (2÷3)[LOG]((5.96×10^16)÷(10^4.40)). The magnitude is 8.25.

Example 9 Textbook Section 4-5

Solve $\ln(x + 3) - \ln x = 5 \ln(x^2 - 4)$ to three decimal places.

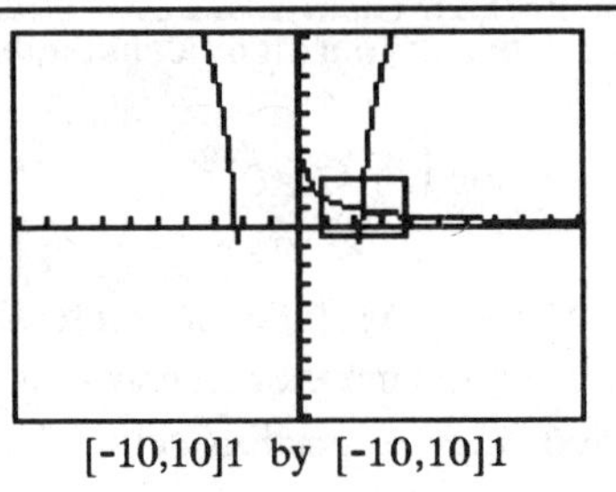

[-10,10]1 by [-10,10]1

Solution: Store each side of the equation as a separate function as: LN (X+3)–LN X and 5 LN (X^2–4). Graph. Use trace and zoom to find the x coordinate of the intersection point. A typical zoom box is shown in the figure to the right. The result is $x \approx 2.277$. Hence the solution to the equation is $x \approx 2.277$. Check this by storing 2.277 as X and evaluating both expressions to three decimal places. The results should be almost the same.

▲▲▲ EXERCISE 4

1. Make a table of values for the functions in Problems 1-10 of textbook Exercise 4-1. Use at least 5 values for x.
2. Graph Problems 1 and 3 of textbook Exercise 4-1. Describe how these functions are related to each other.
3. Graph Problems 2 and 4 of textbook Exercise 4-1. Describe how these functions are related to each other.
4. Graph Problems 1 and 2 of textbook Exercise 4-1. Describe how these functions are related to each other.
5. Graph Problems 3 and 5 of textbook Exercise 4-1. Describe how these functions are related to each other.
6. Solve Problems 17-38 of textbook Exercise 4-1 using graphing techniques.
7. Graph Problems 13-20 of textbook Exercise 4-2.
8. Solve Problems 27-30 of textbook Exercise 4-2 using graphing techniques.
9. Solve Problems 78, 80, and 81 of textbook Exercise 4-3 using graphing techniques.
10. Solve Problem 85-86 of textbook Exercise 4-3 using: (A) $b = 10$ (B) $b = e$
11. Solve Problems 1-24 of textbook Exercise 4-4 using the graphing calculator.
12. Solve Problems 25-32 of textbook Exercise 4-4 using the graphing calculator.
13. Solve Problems 43-46 of textbook Exercise 4-4 using graphing techniques.
14. Solve Problems 19-38 of textbook Exercise 4-5 using graphing techniques.
15. Graph Problems 53-56 of textbook Exercise 4-5. (See Appendix Section A-17, B-17 or C-17 of this manual.)

CHAPTER 5

TRIGONOMETRIC FUNCTIONS

Example 1 Textbook Section 5-2

Evaluate to three decimal place accuracy: (A) $\sin(17\pi/6)$ (B) $\tan(-23\pi/7)$

Solution (A): Enter [SIN] (17 π ÷ 6) into the calculator and evaluate. The result is .500.

Solution (B): Enter [TAN] (⁻ 23 π ÷ 7) into the calculator and evaluate. The result is -1.254.

Example 2 Textbook Section 5-2

Verify the identity $\dfrac{1-\sin^2 x}{1-\cos^2 x} = \dfrac{\cos^2 x}{\sin^2 x}$ to two decimal places for:

(A) $x = .56$ (B) $x = -7.28$ (C) $x = 6.67$ (D) $x = -2.36$

Solution: Set the calculator to radian mode. (See Examples 1 & 2 of Appendix Section A-18, B-18 or C-18 of this manual.) Store the left side of the expression in the calculator as (1-([SIN]X)^2)÷(1-([COS] X)^2). Store the right side of the expression as a second function as (([COS] X)^2)÷(([SIN] X)^2). Store the value of x and evaluate both expressions for each of the angle measures. The results will be the same for both expressions. The results are: (A) 2.54 (B) .42 (C) 6.03 (D) 1.02.

Example 3 Textbook Sections 5-3

Find : (A) The radian measure to three decimal places of an angle of 53°28' 39".

(B) The degree measure to the nearest minute θ_D of the central angle opposite an arc of 53 feet in a circle of radius 21 feet. In which quadrant is the terminal side of this angle.

Solution:

(A) Convert the measure of the angle to decimal degrees by calculating 53+28÷60+39÷3600. The result displayed on the calculator is 53.477. Convert this to radian measure by multiplying by (π/180). Do not reenter the result found in the first calculation. Use the ANS key: [ANS] ×π÷180. The result to three decimal places is .933 radians.

(B) The radian measure is the number of times the radius can be measured along the arc. θ_R = 53÷21 = 2.5238095 radians as displayed on the calculator. The degree measure is found by multiplying the radian measure by 180/π: $\theta_D = \frac{180}{\pi}\ \theta_R$ = (180÷π)× [ANS] =

144.603634 ° as displayed on your calculator. The decimal portion of this degree measure must be converted to minutes and seconds. The steps to find degrees, minutes and seconds for this angle are:

144.603634 - 144	Subtract the degrees.	Result: .603634
[ANS] × 60	This is the decimal minutes.	Result: 36.2180455
[ANS] - 36	Subtract the minutes.	Result: .218040552
[ANS] × 60	This is the decimal seconds.	Result: 13.08243312

The degree measure to the nearest minute is 144° 36' 13".
The terminal side of this angle is in Quadrant II since θ_D is between 90° and 180°.

Example 4 Textbook Section 5-4

Find the radian measure to three decimal places of ß of the right triangle with a = 4.32 cm and b = 2.62 cm.

b = 2.62
ß
a = 4.32

Solution: The radian measure of angle ß is $\tan^{-1}\left(\frac{2.62}{4.32}\right)$.

Set the calculator to radian measure. (See Appendix Section A-18, B-18 or C-18 of this manual.) Enter [TAN^{-1}] [(] [2.62] [÷] [4.32] [)]. The result is: ß ≈ .545 radians.

Example 5 Textbook Section 5-4

Evaluate to 4 decimal places: (A) cot(-143°32') (B) cos(-357.32 radians)

Solution: (A) Set the calculator to degree mode. (See Examples 1 & 2 of Appendix Section A-18, B-18 or C-18 of this manual.) Convert the angle to decimal degrees by calculating -(143 + 32÷60). The result is -143.5333. Now evaluate the cotangent of this angle by entering [1] [÷] [TAN] [ANS]. The result is 1.3531.

(B) Set the calculator to radian mode. (See Appendix Section A-18, B-18, or C-18 of this manual.) Evaluate the cosine by entering [COS] [(-)] [357.32]. The result is .6811.

Example 6 Textbook Section 5-4

Angle θ has terminal side in Quadrant IV as shown in the diagram at the right. Find the reference angle A in degree measure to the nearest minute. Also find the smallest positive angle having the same cosecant.

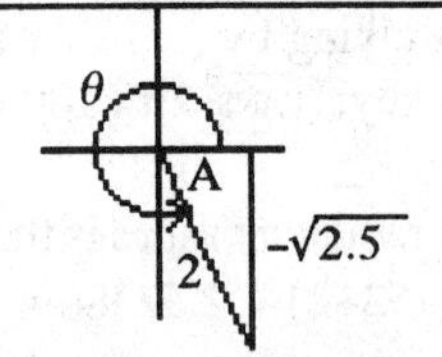

Solution: Set the calculator to degree mode. (See Examples 1 & 2 of Appendix Section A-18, B-18, or C-18 of this manual.) From the diagram we see that the measure of reference angle A can be found using $A = \sin^{-1}\left(\frac{\sqrt{2.5}}{2}\right)$. Evaluate this by entering

[SIN⁻¹] [(] [√] [2.5] [÷] [2] [)]. The result is 52.23875609. Change the decimal part to minutes and seconds using the following sequence of calculations:

[ANS] - 52	Subtract the degrees.	Result: .238756093
[ANS] × 60	Find the decimal minutes.	Result: 14.32536558
[ANS] - 14	Subtract the minutes.	Result: .3253655782
[ANS] × 60	This is the decimal seconds.	Result: 19.52193469

Hence A≈ 52°14'20" . The smallest positive angle having the same cosecant is in Quadrant III. Its measure is 180° + A or 232°14'20" .

Example 7 Textbook Section 5-5

Investigate the behavior of the graph of $y = \tan x$ when x is near $\frac{\pi}{2}$.

Solution: Set the calculator to radian mode. (See Examples 1 & 3 of Appendix Section A-18, B-18 or C-18 of this manual.) Set the RANGE to [-6.28,6.28]1.57 by [-15,15]0. This is approximately [-2π,2π]π/2 by [-15,15]0. The zero means that no scale marks will show on the vertical axis. Graph $y = \tan x$. Use trace and zoom on both sides of $\frac{\pi}{2}$. We see that as x gets close to $\frac{\pi}{2}$ (but less than $\frac{\pi}{2}$), the values of y get larger and larger. The function $\tan x$ appears to increase without bound as x approaches $\frac{\pi}{2}$ from the left. We also see that as x gets close to $\frac{\pi}{2}$ (but a little greater than $\frac{\pi}{2}$), that the values of y get very large in magnitude but are negative. The function $\tan x$ appears to decrease without bound as x approaches $\frac{\pi}{2}$ from the right.

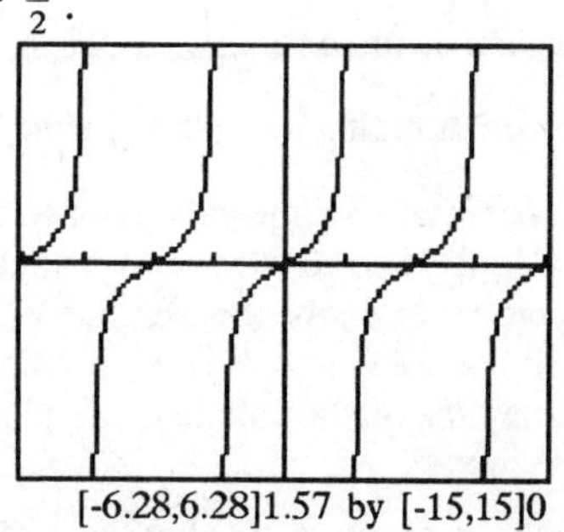
[-6.28,6.28]1.57 by [-15,15]0

Example 8 Textbook Section 5-5

Graph $y = \cos x$ and $y = \sec x$ on the same coordinate axes.

Solution: Set the calculator to radian mode. (See Examples 1 & 3 of Appendix Section A-18, B-18 or C-18 of this manual.) Store $y = \cos x$ and $y = 1/(\cos x)$ as separate functions in the calculator as [cos] and 1/[cos]. Set the RANGE to [-6.28,6.28]1.57 by [-3,3]1. This is approximately [-2π,2π]π/2 by [-3,3]1. Graph. You may see vertical lines on the graph. These lines are not part of the graph nor are they vertical asymptotes. These lines are drawn by the calculator because it is joining one calculated point

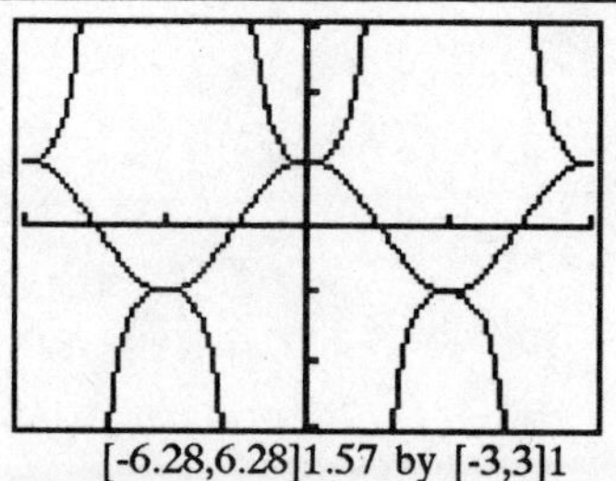
[-6.28,6.28]1.57 by [-3,3]1

with the next calculated point as it uses x values from the Xmin to the Xmax value. The graph will be drawn without these lines by using the dotted mode.

The graph can also be drawn in degree mode. A graph identical to the one shown above will be obtained when setting RANGE [-360,360]30 by [-3,3]1.

Example 9 Textbook Sections 5-6 & 5-7

Compare the graphs of each of the following pairs of functions. Describe the relationship between the functions in each pair:

(A) $f(x) = \sin x$ and $g(x) = 2 \sin x$.

(B) $f(x) = \sin x$ and $h(x) = \sin 2x$

(C) $f(x) = \sin x$ and $k(x) = \sin(x + \pi/2)$

(D) $f(x) = \sin x$ and $m(x) = 2 + \sin x$

(E) Based on the above graphs, describe the graph of $q(x) = -4 - \frac{3}{4}\sin\left(\frac{1}{2}x - \frac{\pi}{3}\right)$ before graphing it? Then graph.

Solution: Set the calculator to radian mode. (See Examples 1 & 3 of Appendix Section A-18, B-18, or C-18 of this manual.) Enter each of the functions in the calculator. Set an appropriate RANGE. See Example 7 above. Graph. The graphs shown at the right illustrate the calculator display. It is not possible to get labeling or to get both dotted and solid graphs at the same time on the calculator display.

(A) The graph of $g(x)$ has the same period but twice the amplitude of the graph of $f(x)$.

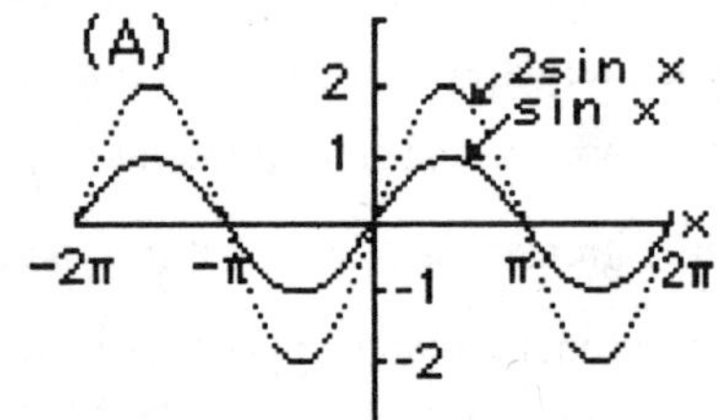

(B) The graph of $h(x)$ has the same amplitude but half the period of the graph of $f(x)$.

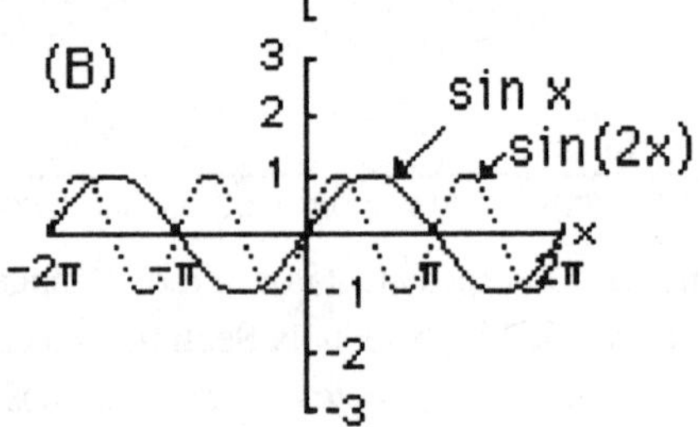

(C) The graph of $k(x)$ is shifted $\pi/2$ units to the left from the graph of $f(x)$.

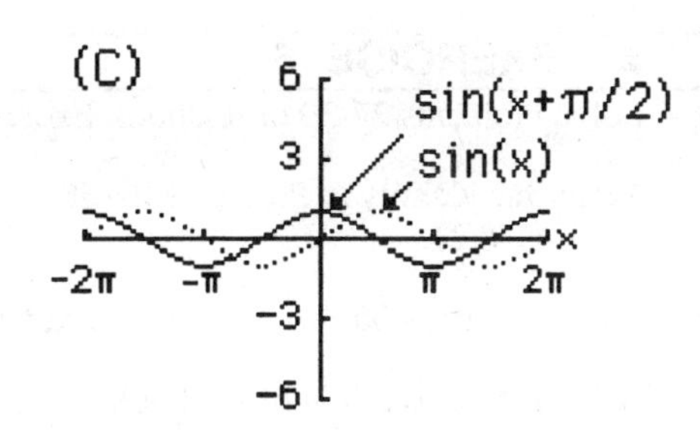

(D) The graph of $m(x)$ has the same period and amplitude but is shifted vertically 2 units from the graph of $f(x)$.

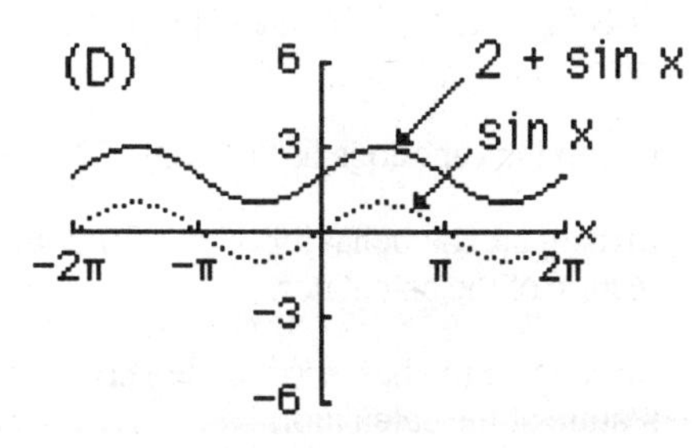

(E) The graph of $q(x)$ will be
shifted down 4 units
have twice the period
will be shifted $\pi/3$ units to the right
will have negative three-fourths the amplitude
of the graph of $f(x) = \sin x$.

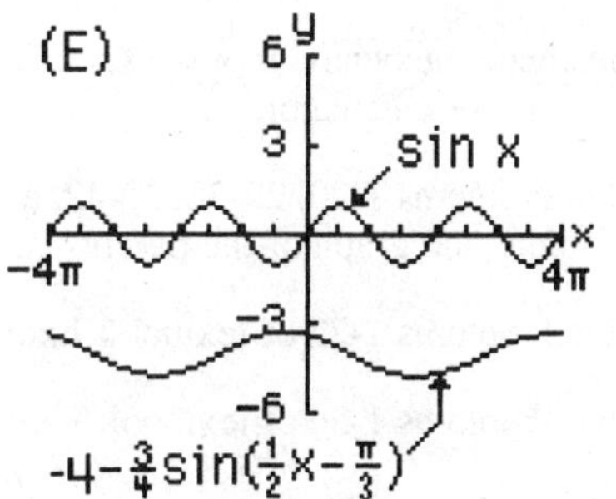

Example 10 Textbook Section 5-8

Graph $y = \cos^{-1}x$ and $y = 2\cos^{-1}3x$ on the same coordinate axes. What changes in the graph occur because of the 2 and 3 in the second function?

Solution: Set the calculator to radian mode. (See Examples 1 & 4 of Appendix Section A-18, B-18 or C-18 of this manual.) Store $y = \cos^{-1}x$ and $y = 2\cos^{-1}3x$ as [cos⁻¹] X and 2 [cos⁻¹] 3 X separately in the calculator. Set the RANGE to [-1,1].5 by [0,6.28]3.14. Graph both functions on the same coordinate axes. The second curve is stretched vertically by a factor of 2 and compressed horizontally by a factor of 3.

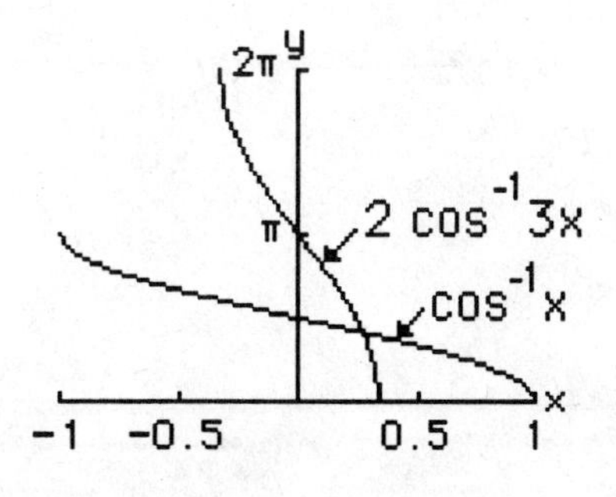

▲▲▲ EXERCISE 5

1. Solve Problems 27-30 of textbook Exercise 5-2.

2. Verify the identity in Problems 87-88 of textbook Exercise 5-2 using $x = -24.7$ and $x = 348.21$.

3. Solve Problems 35-40 of textbook Exercise 5-3 using the calculator.

4. Solve Problems 31-54 of textbook Exercise 5-4 using the calculator.

5. Solve Problems 5-16, and 21-26 of textbook Exercise 5-5 by examining the graph of the function.

6. Graph the functions in Problems 27-36 of textbook Exercise 5-5.

7. Investigate the behavior of $y = \cot x$ when x is near π by graphing and using the TRACE feature of the calculator.

8. Investigate the behavior of $y = \sec x$ when x is near $\pi/2$ by graphing and using the TRACE feature of the calculator.

9. Investigate the behavior of $y = \csc x$ when x is near 0 by graphing and using the TRACE feature of the calculator.

10. Solve Problems 1-20, 29-36, 39-42, and 47-50 of textbook Exercise 5-6 by graphing and examining the graph of the function.

11. Solve Problems 1-22 of textbook Exercise 5-7.

12. Solve Problems 1-54 of textbook Exercise 5-8 using a calculator.

CHAPTER 6

TRIGONOMETRIC IDENTITIES AND CONDITIONAL EQUATIONS

Example 1 Textbook Sections 6-1, 6-2, 6-3, and 6-4

Evaluate both sides of $\frac{1 + \sin x}{\cos x} + \frac{\cos x}{1 + \sin x} = 2 \sec x$ to three decimal places for:
(A) x = 36°42'18" (B) x = -108.95° (C) x = -643.26.

Solution (A): For x = 36°42'18". Set the calculator to degree mode. (See Appendix Section A-18, B-18, or C-18 of this manual.) Store each side of the identity as a separate function: (1+[SIN] X)÷[COS] X + [COS]X ÷(1+[SIN] X) and 2(1÷[COS] X). Change the measure of x to decimal form by calculating 36+42÷60+18÷3600 (see Example 1(A) of Chapter 6 of this manual). The value of x is 36.705. Store this value as x in the calculator by storing [ANS] as x . Evaluate each of the functions. Both of the results should be the same. The value is 2.495 accurate to three decimal places.

Solution (B): For x = -108.95°. Set the calculator to degree mode. (See Appendix Section A-18, B-18, or C-18 of this manual.) Store -108.95 as X in the calculator and evaluate each of the functions. The results should be the same. The value of the functions is -6.159 accurate to three decimal places.

Solution (C): For x = -643.26. Set the calculator to radian mode. (See Appendix Section A-18, B-18, or C-18 of this manual.) Store -643.26 as X in the calculator and evaluate each of the functions. The results should be the same. The value of the functions is -2.776 accurate to three decimal places.

Example 2 Textbook Sections 6-1, 6-2, 6-3, and 6-4

Verify the identity in Example 1 graphically.

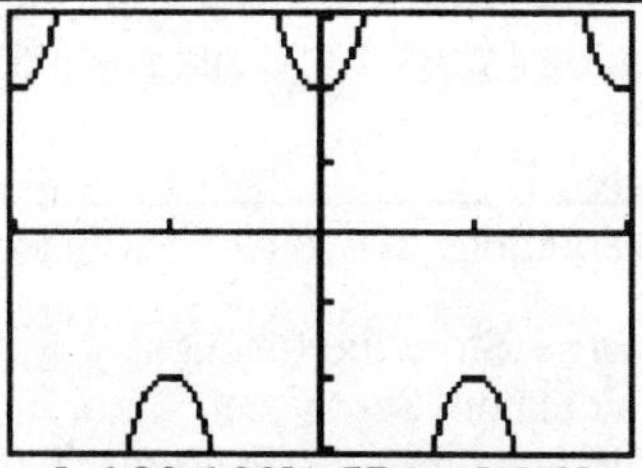

[-6.28,6.28]1.57 by [-3,3]1

Solution: Set the calculator to radian mode. (See Appendix Section A-18, B-18, or C-18 of this manual.) Set the RANGE values to [-6.28,6.28]1.57 by [-3,3]1. Store each side of the identity as a separate function as in Example 1 above. (See Appendix Section A-13, B-13 or C-13 of this manual.) Graph both of the functions. Note that when graphing both functions on the same coordinate axes, the function is displayed but then a time passes when nothing seems to be happening. What is actually happening is that the second graph is being drawn exactly over the first graph.

This may seem to verify that the identity is true since they appear the same for all values of x. In actuality, the only way to verify an identity is algebraically. Why? Graph $y = 1.001(2\sec x)$, $y = 2.004 \sec x$, $y = 2 \sec 1.001x$ and $y = 2 \sec x$ on the same axes. The graphs are exactly as that shown above. Does this mean the functions are identical? No! The graphs look alike because of the rounding done by the calculator and the way functions are graphed by the calculator.

What happens if the calculator is set in degree mode instead of radian mode? You will need to use [-360,360]90 by [-3,3]1 to see the graph. The graphs will again look alike.

Example 3 — Textbook Section 6-5

Find all solutions to three decimal place accuracy of each of the following for $0 \le x \le 2\pi$:

(A) $\cos x = 0.87$ (B) $\cot\left(\frac{\pi}{2} - x\right) = 15.83$

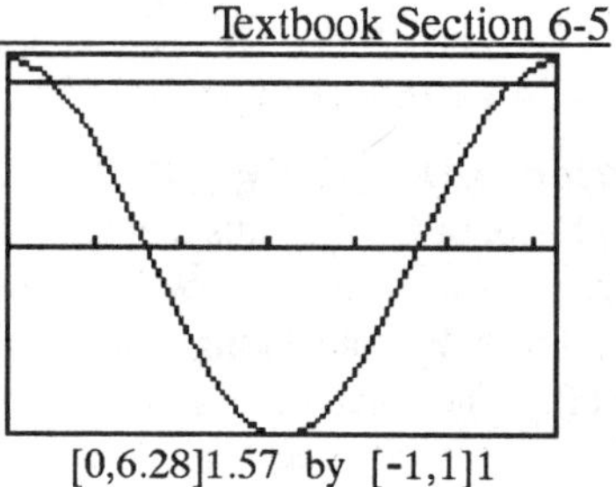

[0,6.28]1.57 by [-1,1]1

Solution: (A) Store the functions $y = \cos x$ and $y = .87$ separately in the calculator. Set the calculator to radian mode. (See Appendix Section A-18, B-18, or C-18 of this manual.) Set the RANGE to [0,6.28]1.57 by [-1,1]1. This was chosen because we are given the RANGE for the x variable and we know $-1 \le \cos x \le 1$. Graph both functions on the same coordinate axes. Use trace and zoom to find the x value of the intersection points. The results are .516 and 5.768 The solution to the equation is $x \approx .516$ and $x \approx 5.768$.

(B) Store the functions $y = 1/\tan\left(\frac{\pi}{2} - x\right)$ and $y = 15.83$ in the calculator. Set the calculator in radian mode. (See Appendix Section A-18, B-18, or C-18 of this manual.) Set the RANGE to [0,6.28].785 by [0,20]2. Graph and find the x value of the two intersection points. They are $x \approx 1.508$ and $x \approx 4.649$. Hence the solutions to the equation are $x \approx 1.508$ and $x \approx 4.649$.

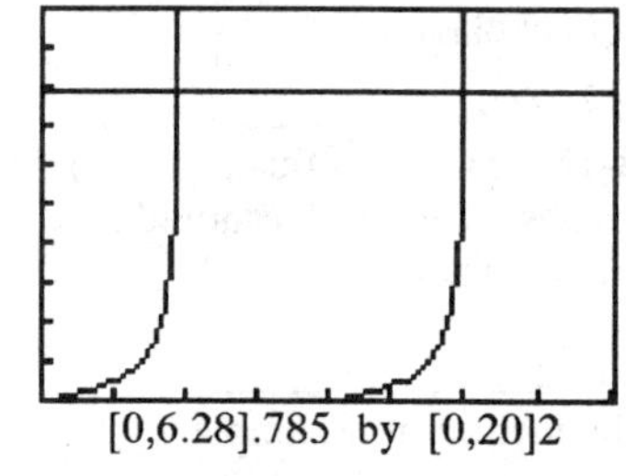

[0,6.28].785 by [0,20]2

Example 4 — Textbook Section 6-5

Find all solutions to two decimal places for $\sec x = 2.43$.

Solution: Store the functions $y = 1/(\cos x)$ and $y = 2.43$ in the calculator. Set the calculator in radian mode. (See Appendix Section A-18, B-18, or C-18 of this manual.) Set the RANGE to [-6.28,6.28]3.14 by [0,3]1 . Graph the functions. Use trace and zoom to find the x value of one of the intersection points. A typical zoom box is shown in the figure at the right. The point indicated on the graph has $x \approx 1.15$. The solution set can be found from this single value. It is $\{x \mid x \approx 1.15 \pm 2\pi, x \approx -1.15 \pm 2\pi\}$

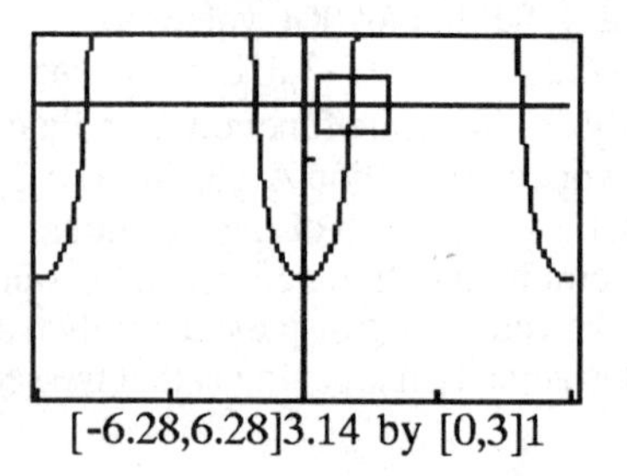

[-6.28,6.28]3.14 by [0,3]1

Example 5 Textbook Section 6-6

Find solutions to three decimal places for
2 tan $3x$ - 5.87 = 2 sin $2x$ for $0 \le x \le \pi/2$.

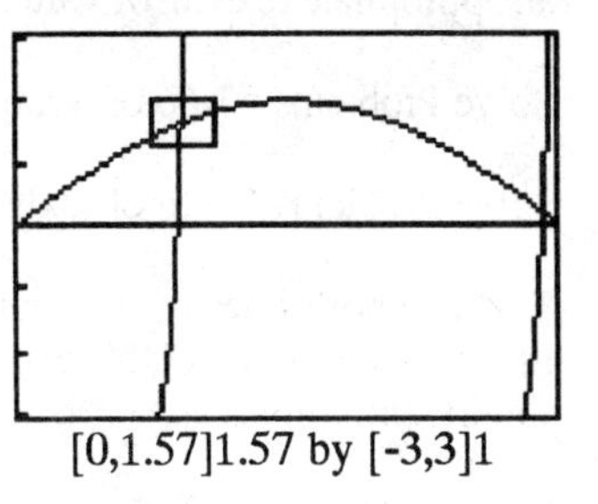

[0,1.57]1.57 by [-3,3]1

Solution: Set the calculator to radian mode. (See Appendix Section A-18, B-18, or C-18 of this manual.) Set the RANGE to [0,1.57]1.57 by [-3,3]1 . Store the left side of the equation and the right side of the equation as separate functions in the calculator as 2 [TAN] (3X)−5.87 and 2 [SIN] (2X). Graph. Use trace and zoom to find the x value of the two intersection points. A typical zoom box is shown for one of the intersection points in the figure at the right. The x values are the solutions to the equation. They are: $x \approx .436$ and $x \approx 1.468$.

▲▲▲ EXERCISE 6

1. Evaluate both sides of the expressions in Problems 1-76 and 85-87 of textbook Exercise 6-1 using angle measures of 18°35'9", -5899.43°, and 219.57 radians. Evaluate each side to 5 significant digits.

2. Graph both sides of the expressions in Problems 1-76 and 85-87 of textbook Exercise 6-1 as separate functions using $[-2\pi,2\pi]\pi/2$ for the x values and an appropriate RANGE for the y values.

3. Evaluate both sides of the expressions in Problems 1-4, 9-12, and 31-42 of textbook Exercise 6-2 using the angle measure of 35.67 radians. Use -257.8 radians for the second angle in problems requiring two angle measures. Evaluate each side to 5 significant digits.

4. Graph both sides of the expressions in Problems 1-4, 9-12, 31-32, and 35-36 of textbook Exercise 6-2 using $[-2\pi,2\pi]\pi/2$ for the x values and an appropriate RANGE for the y values.

5. Solve Problems 45-52 of textbook Exercise 6-2 using the calculator.

6. Solve Problems 1-10 of textbook Exercise 6-3.

7. Evaluate both sides of the expressions in Problems 11-34 of textbook Exercise 6-3 using an angle measure of 35.892°. Evaluate each side to 5 significant digits.

8. Graph both sides of the expressions in Problems 11-34 of textbook Exercise 6-3. Choose an appropriate RANGE of values for the graphing screen.

9. Solve Problems 53-66 of textbook Exercise 6-3.

10. Solve Problems 10-16 of textbook Exercise 6-4 using a calculator.

11. Evaluate both sides of the expressions in Problems 17-33 of textbook Exercise 6-4 using 81.74 for the first variable and 381.9 for the second variable. Evaluate each side to 5 significant digits.

12. Solve Problems 1-62 of textbook Exercise 6-5 using the calculator and the technique illustrated in Examples 3-5 of this chapter of the manual.

13. Solve Problems 1-58 of textbook Exercise 6-6 using graphing techniques.

CHAPTER 7

ADDITIONAL TOPICS IN TRIGONOMETRY

Example 1 Textbook Section 7-2

Find side a to three decimal place accuracy in the triangle shown at the right.

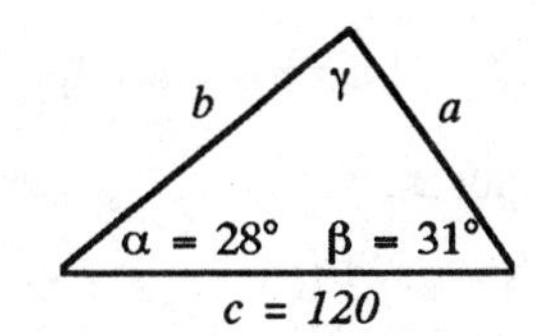

Solution: Using the law of sines we have

$$a = \frac{c \sin\alpha}{\sin\gamma} = \frac{120 \sin 28°}{\sin 121°} \text{ where}$$

$\gamma = 180° - (28° + 31°) = 121°.$

Set the calculator to degree mode and enter the expression 120 × [SIN] 28 ÷ [SIN] 121 and execute. The result is 65.724. So $a \approx 65.724$.

Another way to solve this is to store the expression C sin A ÷ sin G as a function. Then store 120 as C, 28 as A, and 121 as G, and then evaluate the expression.

Example 2 Textbook Section 7-3

Find side a to two decimal place accuracy in the triangle shown at the right.

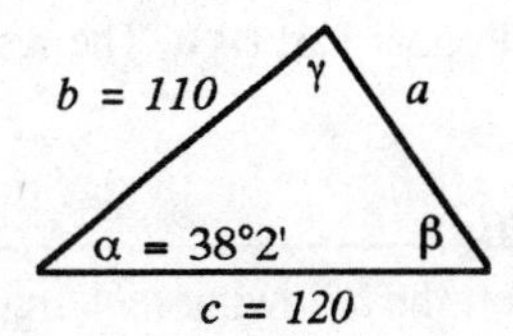

Solution: Set the calculator to degree mode. Using the law of cosines we have $a = \sqrt{b^2 + c^2 - 2bc\cos\alpha}$.

Store the right side of this in the calculator as

√ (B ^ 2 + C ^ 2 - 2 B C [COS] A).

Store 110 as B, 120 as C, and 38 + 2 ÷ 60 as A. Evaluate the expression. The result is $a \approx 75.54$.

Example 3 Textbook Section 7-4

Find, to two decimal places, the magnitude of the resultant vector **u + v** and α for each of the following:

(A) $|\mathbf{u}| = 32$, $|\mathbf{v}| = 15$, $\theta = 68°$
(B) $|\mathbf{u}| = 120$, $|\mathbf{v}| = 90$, $\theta = 10°$
(C) $|\mathbf{u}| = 28$, $|\mathbf{v}| = 56$, $\theta = 34.58°$

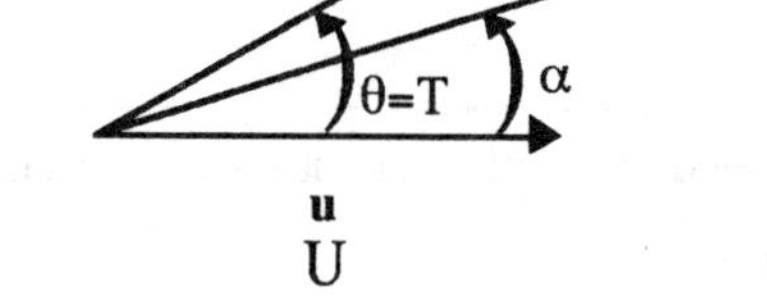

Solution:
Step 1 Store the expression

$$\sqrt{|u|^2 + |v|^2 - 2|u||v|\cos(180\text{-}\theta)}$$

in the calculator as

√ (U ^ 2 + V ^ 2 - 2 U V [COS] (180 - T)).
U represents $|\mathbf{u}|$ and V represents $|\mathbf{v}|$.

Step 2 Store the expression $\sin^{-1}\left(\frac{V \sin(180\text{-}T)}{W}\right)$ in the calculator as

[SIN⁻¹] ((V [SIN] (180 - T)) ÷ W) .

Step 3
(A) Store 32 as U, 15 as V, and 68 as T.
Set the calculator to degree mode and evaluate the first expression. The result is 40.11....
Store [ANS] as W . Evaluate the second expression. The result is 20.29. So
$\alpha = 20.29°$.

(B) Store 120 as U, 90 as V and 10 as T.
Repeat Part (A). The result is 4.28. So $\alpha = 4.28°$.

(C) Store 28 as U, 56 as V and 34.58 as T.
Repeat Part (A). The result is 23.21. So $\alpha = 23.21°$.

Example 4 Textbook Section 7-6
Convert the following rectangular coordinates to polar coordinates where $0 \le \theta \le 2\pi$.

(A) (-3, 56) (B) (56, -23) (C) (-4, -7)

Solution: One way to do this is to use the built-in functions of the calculator. (See Section A-19, B-19 and C-19 of this manual on how to use the special calculator functions to make this conversion.) First set the calculator to radian mode since the measure of θ is in radians.

The answers are: (A) (56.08, 1.62) (B) (60.54, -.39) (C) (8.06, -2.09)

Example 5 Textbook Section 7-6

Convert the following polar coordinates to rectangular coordinates:

(A) (-3, 56°) (B) (38, -54°) (C) (-3, -45)

Solution: One way to do this is to use the built-in functions of the calculator. (See Section A-19, B-19 and C-19 of this manual on how to use the special calculator functions to make this conversion.)

(A) Set the calculator to degree mode. The rectangular coordinates are: (-1.678, -2.487)
(B) Set the calculator to degree mode. The rectangular coordinates are: (22.336, -30.743)
(C) Set the calculator to radian mode. The rectangular coordinates are: (-1.576, 2.553)

Example 6 Textbook Section 7-7

Graph $r = 4\theta$ for $0 \le \theta \le 4\pi$.

Solution: The largest value for r will be $4 \times 4\pi$ or approximately 50.2. Hence, the RANGE values for the graphing screen could be [-60,60]5 by [-60,60]5. However the graph will look better if the scale marks along both axes have the same spacing. Hence use [-90,90]5 by [-60,60]5.

[-90,90]5 by [-60,60]5

Note that the RANGE for y values is 2/3 the RANGE for the *x* values. This will give equal distance between scale marks on the axes.

The graphing procedures for polar coordinates are discussed in Appendix Sections A-19, B-19, and C-19 of this manual.

Example 7 Textbook Section 7-7

Graph $r = 6 \sin 3\theta$ for $0 \le \theta \le \pi$.

Solution: The largest value for $\sin 3\theta$ is 1. So the largest value for r will be 6. The RANGE values could be [-6,6]1 by [-6,6]1. However the graph will look better if the scale marks along both axes have the same spacing. Hence use [-9,9]1 by [-6,6]1.

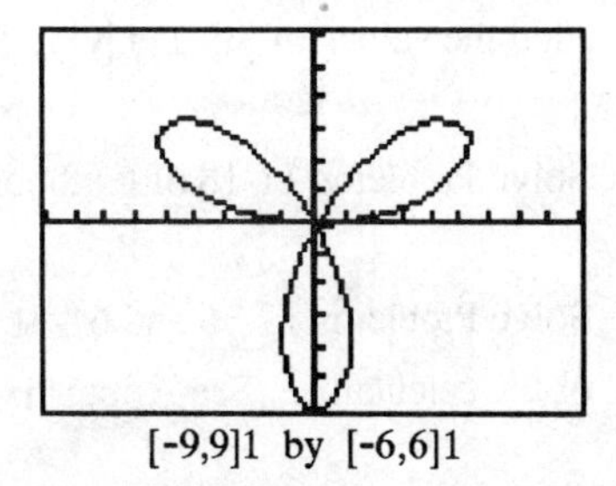

[-9,9]1 by [-6,6]1

Note the RANGE for y values is 2/3 the RANGE for the x values. This will give equal distance between scale marks on the axes.

The graphing procedures for polar coordinates are discussed in Appendix Sections A-19, B-19, and C-19 of this manual.

▲▲▲ EXERCISE 7

1. Solve Problems 1-14 of textbook Exercise 7-1 using the calculator.

2. Store two forms of law of sines in the calculator as two separate functions:
   (A) to solve for side a of a triangle store (B [SIN] H) ÷ [SIN] D where H represents α, B represents b, and D represents β.
   (B) to solve for an angle β of a triangle store [SIN^{-1}] ((B [SIN] H) ÷ A) where H represents α, B represents b, and A represents a.

   Solve Problems 13-40 of textbook Exercise 7-2 by storing the values given in the problem and evaluating the appropriate stored expression.

3. Store the left side and the right side of Mollweide's equation in Problem 43 of textbook Exercise 7-1 as two separate functions in the calculator. Use this to check your solution to Problems 13-40 of textbook Exercise 7-2.

4. Store the law of sines as two expressions given in Problem 2 above and the law of cosines as a third function in the calculator as: √(B ^ 2 + C ^ 2 - BC [COS] H)
   Use these stored functions to solve Problems 1-24 of textbook Exercise 7-3.

5. Store the expression [TAN^{-1}] (V ÷ U) where U represents the vector length $|\mathbf{u}|$ and V represents the vector length $|\mathbf{v}|$. Use this to find the value of θ in Problems 1-6 of textbook Exercise 7-4 by storing the values of U and V and evaluating the expression.

6. Find the values of $|\mathbf{u}|$ and $|\mathbf{v}|$ in Problems 7-10 of textbook Exercise 7-4.

7. Solve Problems 11-18 of textbook Exercise 7-4 by first storing an appropriate expression.

8. Solve Problems 17-36 and 61-71 of textbook Exercise 7-6 by using the special functions of the calculator. (See Appendix Sections A-19, B-19, or C-19 of this manual.)

9. Graph Problems 1-30 of textbook Exercise 7-7.

CHAPTER 8

SYSTEMS OF EQUATIONS AND INEQUALITIES

Example 1 Textbook Section 8-1 and 8-2

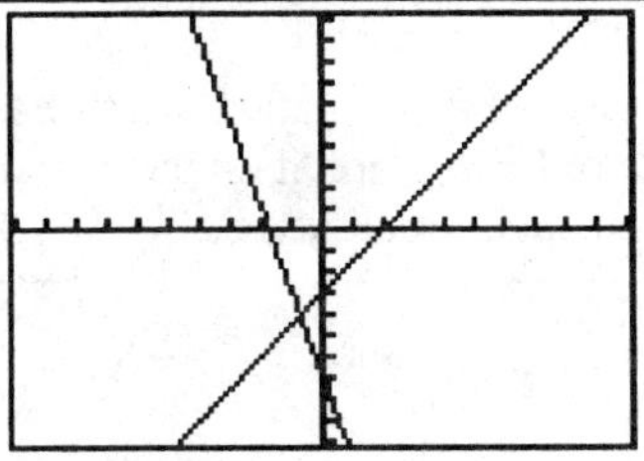

Solve the system to three decimal place accuracy using graphing and augmented matrix methods:

$$3x_1 - 2x_2 = 6$$
$$4x_1 + x_2 = -7$$

Solution: The equations must be written with x as x_1 and y as x_2:

[-10,10]1 by [-10,10]1

$$3x - 2y = 6$$
$$4x + y = -7$$

Also, each equation must be solved for y in order to enter it into the calculator:

$$y = (6-3x)/(-2)$$
$$y = -7-4x$$

Store these in the calculator as two separate functions: (6-3X)÷(-2) and -7-4X.
Graph using [-10,10]1 by [-10,10]1 to get the graph shown above. Use trace and zoom to find the intersection point. The result is (-.727,-4.091). Hence the solution to the original system is $x_1 \approx -.727$ and $x_2 \approx -4.091$.

The augmented matrix for this system is:

$$\left[\begin{array}{cc|c} 3 & -2 & 6 \\ 4 & 1 & -7 \end{array}\right]$$

Enter this as a 2x3 matrix in the calculator. (See Appendix Section A-15, B-15, or C-15 of this manual.) Then perform the following row operations:

$$\underset{\sim}{R_1 \leftrightarrow R_2} \left[\begin{array}{cc|c} 4 & 1 & -7 \\ 3 & -2 & 6 \end{array}\right] \underset{\sim}{\tfrac{1}{4}R_1 \to R_1} \left[\begin{array}{cc|c} 1 & .25 & -1.75 \\ 3 & -2 & 6 \end{array}\right]$$

$$\underset{\sim}{R_2 - 3R_1 \to R_2} \left[\begin{array}{cc|c} 1 & .25 & -1.75 \\ 0 & -2.75 & 11.25 \end{array}\right] \underset{\sim}{\tfrac{1}{-2.75}R_2 \to R_2} \left[\begin{array}{cc|c} 1 & .25 & -1.75 \\ 0 & 1 & -4.090\ldots \end{array}\right]$$

$$R_1 + (-.25)R_2 \to R_1 \quad \sim \quad \left[\begin{array}{cc|c} 1 & 0 & -.727\ldots \\ 0 & 1 & -4.090\ldots \end{array}\right]$$

The corresponding system is $x_1 = -.727$ and $x_2 = -4.091$ rounded to three decimal places.

Example 2 Textbook Section 8-3

Find the real solutions to two decimal place accuracy of the system:

$x^2 - 2y^2 = 2$

$xy = 3$

Solution: Each equation must be solved for y in order to graph it on the calculator. The first equation results in two functions:

$$y = \sqrt{\frac{2-x^2}{-2}} \quad \text{and} \quad y = -\sqrt{\frac{2-x^2}{-2}}$$

The second equation is $y = \dfrac{3}{x}$.

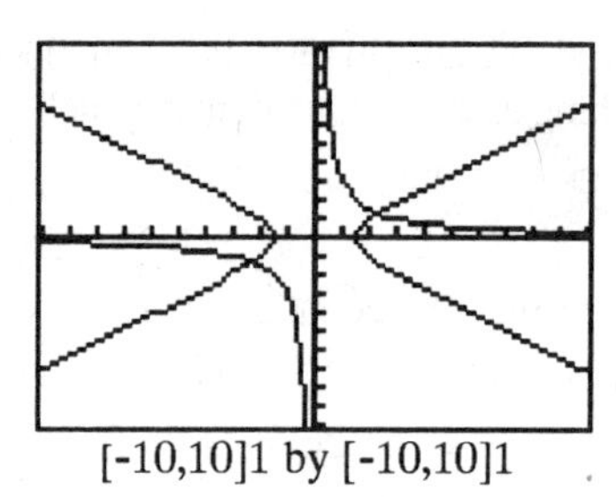

[-10,10]1 by [-10,10]1

Enter these as three separate functions into the calculator as:

√((2 - X^2) ÷ - 2), -√((2 - X^2) ÷ - 2), and 3 ÷ X.

Graph using [-10,10]1 by [-10,10]1 to get the graph shown above.
Use trace and zoom to find the intersection points. They are approximately (2.31, 1.30) and (-2.31, -1.30). The solutions to this system are $x \approx 2.31$, $y \approx 1.30$ and $x \approx -2.31$, $y \approx -1.30$.

Example 3 Textbook Section 8-4

Solve the system of inequalities graphically:

$x + 3y \geq 5$

$2x - y < 4$

Solution:

Step 1. Graph $x + 3y = 5$ and $2x - y = 4$ on the same set of coordinate axes.

Each of these equations must be solved for y in order to enter them into the calculator. They are:

$$y = \frac{5 - x}{3} \quad \text{and } y = 2x - 4.$$

Enter these into the calculator as two separate functions as:

(5 - X) ÷ 3 and 2 X - 4

Graph using [-10,10]1 by [-10,10]1.
See Figure (A) to the right.

(A)

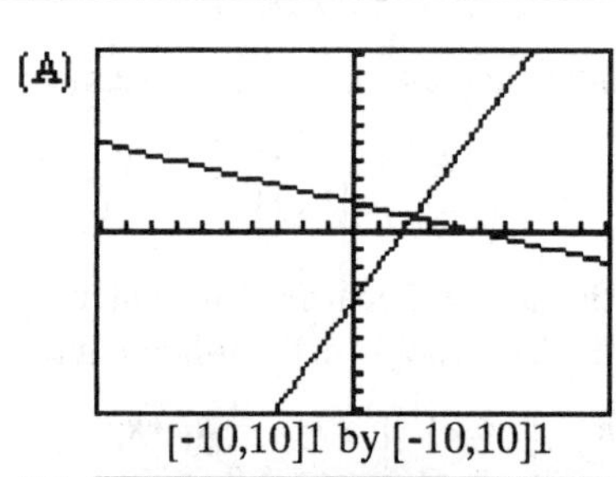

[-10,10]1 by [-10,10]1

(B)

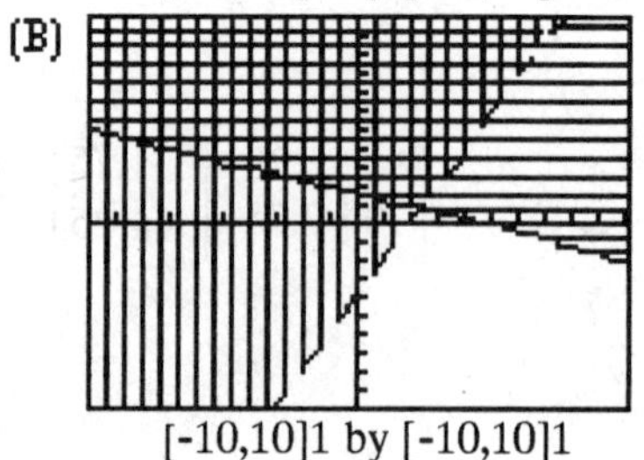

[-10,10]1 by [-10,10]1

Step 2. Pick a point not on either line. The origin (0,0) is not on either line. Substituting x=0 and y=0 into each of the inequalities produces a false statement for the first inequality and a true statement for the second inequality.

Step 3. Using the result from *Step 2* we see that the solution to the system of inequalities is the intersection of the upper half-plane containing the line $x + 3y = 5$ but not containing the point (0,0) (horizontal shading) and the upper half-plane containing the origin but not the line $y = 2x - 4$ (vertical shading). See Figure (B) above. (The shading and dashed line as shown are not possible on the calculator. These features must be added to the graph after it is transferred by hand to paper.) Hence the solution to the system of inequalities is the cross hatched area including the line $x + 3y = 5$ but not including the line $2x - y = 4$.

Example 4 Textbook Section 8-5

Minimize and maximize $z = 3x + 8y$
Subject to

$$x + 3y \le 55$$
$$x + y \ge 10$$
$$x - y \ge 0$$
$$x, y \ge 0$$

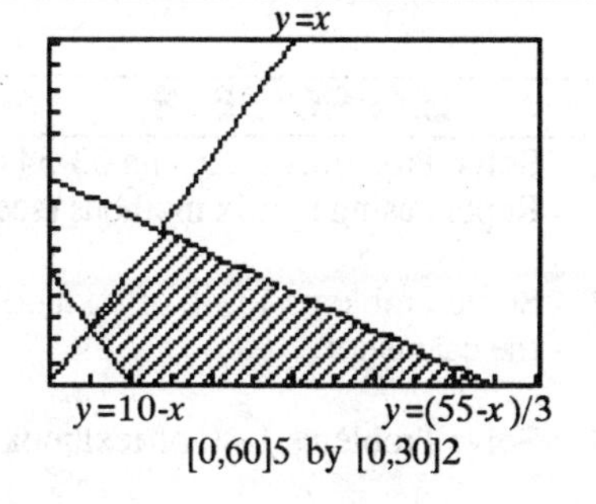

[0,60]5 by [0,30]2

Solution:

Graph the feasible region.

Step 1. Graph the lines $x + 3y = 55$, $x + y = 10$, and $x - y = 0$. In order to enter these into the calculator they must be written as:

$$y = \frac{55 - x}{3}, \quad y = 10 - x, \quad \text{and} \quad y = x.$$

Store these separately in the calculator as: (55 - X) ÷ 3, 10 - X, and X.
Graph using [0,60]5 by [0,30]2. We need only graph the first quadrant because of the nonnegativity constraints: $x \ge 0$ and $y \ge 0$.

Step 2. Determine the feasible region. This can be done using a single point if it is not on any of the lines. Evaluate each of the inequalities at the chosen point. If the inequality is true for this point then the point is in the feasible region and we want the half-plane in which the point lies. Using (1,0) we see that the first and third inequalities are true whereas the second is false.

Step 3. Using the results from *Step 2* we see that the feasible solution region is as shown above. The shading must be added after the graph is copied to paper by hand .

Find the corner points

Use trace and zoom to find the intersection of the lines. The corner points are: (13.75, 13.75), (5, 5), (10, 0), and (55, 0) accurate to two decimal places.

Substitute the corner points into the objective function
Store the function 3X + 8Y in the calculator. Store 13.75 as X and 13.75 as Y and evaluate the objective function. Repeat using the other corner points. The results are given below:

Corner Point	Objective Function
(13.75,13.75)	151.25
(5,5)	55
(10,0)	30
(55,0)	165

The maximum value of the objective function is 165. It occurs when $x=55$ and $y=0$.
The minimum value of the objective function is 30. It occurs when $x=10$ and $y=0$.

▲▲▲ EXERCISE 8

1. Solve Problems 1-20, and 33-54 of textbook Exercise 8-1 using graphing methods. Repeat using matrix methods (see Example 1 above).

2. Solve Problems 17-46 of textbook Exercise 8-2 using matrix row reduction operations on the calculator.

3. Solve Problems 1-40 of textbook Exercise 8-3 using graphing methods.

4. Solve Problems 1-10 and 15-42 of textbook Exercise 8-4.

5. Solve Problems 9-20 of textbook Exercise 8-5.

CHAPTER 9

MATRICES AND DETERMINANTS

Example 1 Textbook Section 9-1

Calculate $A + 3B$ for:

$$A = \begin{bmatrix} 2 & -3 & 0 \\ 1 & -2 & -5 \end{bmatrix} \quad B = \begin{bmatrix} -5 & 6 & 0 \\ 3 & -8 & 7 \end{bmatrix}$$

Solution: Store the matrices in the calculator as [A] and [B]. (See Appendix Section A-15, B-15, or C-15 of this manual.) Calculate $A+3B$ using the matrix multiplication and addition features of the calculator. The result is:

$$A+3B = \begin{bmatrix} -13 & 15 & 0 \\ 10 & -26 & 16 \end{bmatrix}$$

Example 2 Textbook Section 9-2

Calculate AB for:

$$A = \begin{bmatrix} -2 & 4 \\ 2 & 7 \\ 4 & 0 \end{bmatrix} \quad B = \begin{bmatrix} -5 & 0 & 6 \\ 2 & 8 & 9 \end{bmatrix}.$$

Solution: Store the matrices in the calculator as [A] and [B]. (See Appendix Section A-15, B-15, or C-15 of this manual.) Calculate AB using the matrix multiplication features of the calculator. The result is:

$$AB = \begin{bmatrix} 18 & 32 & 24 \\ 4 & 56 & 75 \\ -20 & 0 & 24 \end{bmatrix}$$

Example 3 Textbook Section 9-3

Given M, find M^{-1} if it exists. Round numbers in answer to three decimal places.

$$M = \begin{bmatrix} -5 & 6 & 0 & -2 & 3 \\ 15 & 7 & -11 & 8 & 0 \\ -23 & 5 & 1 & 7 & 0 \\ 9 & 5 & 3 & 6 & -4 \\ 0 & -1 & 6 & 2 & 0 \end{bmatrix}$$

Solution: Store the matrix in the calculator as A. (See Appendix Section A-15, B-15, or C-15 of this manual.) Calculate A^{-1} using the matrix inverse feature of the calculator. The result rounded to three decimal places is:

$$\begin{bmatrix} .008 & .017 & -.032 & .006 & .034 \\ .107 & -.015 & -.002 & .081 & -.068 \\ .037 & -.026 & -.014 & .028 & .107 \\ -.057 & .071 & .042 & -.043 & .144 \\ .093 & .106 & -.022 & -.180 & .287 \end{bmatrix}$$

Example 4 Textbook Section 9-4

Solve the system of equations using inverses. Approximate the solution to three decimal places.

$$\begin{aligned} 2x_1 + 3x_2 - 5x_3 + x_4 - 3x_5 &= 3 \\ x_1 - 4x_2 + 6x_3 - x_4 - 2x_5 &= 8 \\ 3x_1 - 2x_2 + 7x_3 + x_4 + 3x_5 &= -2 \\ x_1 + x_2 - x_3 + 2x_4 + x_5 &= 9 \\ -x_1 + 6x_2 + 8x_3 + 2x_4 + 6x_5 &= 0 \end{aligned}$$

Solution: Let A be the coefficient matrix. Store A in the calculator and find A^{-1} using the matrix inverse feature of the calculator. (See Appendix Section A-15, B-15, or C-15 of this manual.) Store the result as a matrix in the calculator.

$$A^{-1} = \begin{bmatrix} 213... & -.164... & .321... & -.289... & -.060... \\ .158... & -.036... & .004... & -.197... & .097... \\ .015... & .131... & -.034... & .007... & .067... \\ -.128... & .298... & -.273... & .813... & .036... \\ .100... & -.265... & .185... & -.132... & -.042... \end{bmatrix}$$

Store the constant column matrix as B. $B = \begin{bmatrix} 3 \\ 8 \\ -2 \\ 9 \\ 0 \end{bmatrix}$

Now find the product $A^{-1}B$. Do not round numbers and reenter into the calculator. Use A^{-1} as it is stored in the calculator.

The result is $\begin{bmatrix} -3.922 \\ -1.605 \\ 1.234 \\ 9.873 \\ -3.985 \end{bmatrix}$.

The solution to the system is $x_1 = -3.922$, $x_2 = -1.605$, $x_3 = 1.234$, $x_4 = 9.873$, and $x_5 = -3.985$. The solution has been rounded to three decimal places.

Example 5 Textbook Section 9-5 and 9-7

Find the value to three decimal places of x_1 in the system of equations in Example 4 above using Cramer's rule.

Solution: Store the coefficient matrix in the calculator and find the determinant of it using the built-in features of the calculator. (See Appendix Section A-15, B-15, or C-15 of this manual.) Store the result as a variable D. Now change the values in the first column of this matrix to those of the column matrix of constants in the system. Find the value of this determinant. Store this result as another variable say A. Calculate A÷D to find the value of x_1.

$$D = \begin{vmatrix} 2 & 3 & -5 & 1 & -3 \\ 1 & -4 & 6 & -1 & -2 \\ 3 & -2 & 7 & 1 & 3 \\ 1 & 1 & -1 & 2 & 1 \\ -1 & 6 & 8 & 2 & 6 \end{vmatrix} = 1640 \quad \text{and} \quad A = \begin{vmatrix} 3 & 3 & -5 & 1 & -3 \\ 8 & -4 & 6 & -1 & -2 \\ -2 & -2 & 7 & 1 & 3 \\ 9 & 1 & -1 & 2 & 1 \\ 0 & 6 & 8 & 2 & 6 \end{vmatrix} = -6432$$

So $x_1 = \dfrac{A}{D} = -3.922$.

▲▲▲ EXERCISE 9

1. Solve Problems 21-34 and 43-46 of textbook Exercise 9-1.
2. Solve Problems 1-38 of textbook Exercise 9-2.
3. Solve Problems 1-36 of textbook Exercise 9-3.
4. Solve Problems 9-20 of textbook Exercise 9-4.
5. Solve Problems 1-6, 23-28 and 33-44 of textbook Exercise 9-5 using the calculator to evaluate the determinant.
6. Solve Problems 1-28 of textbook Exercise 9-7.

NOTES

CHAPTER 10

SEQUENCES AND SERIES

Example 1 Textbook Section 10-1, 10-2, and 10-3

Evaluate $\sum_{k=1}^{5} \frac{k^3-2k}{k+1}$ to three decimal place accuracy.

Solution: Change the k to x and store the function as: (X ^ 3 - 2 X) ÷ (X + 1).

Store the value of 1 as X and evaluate the function at 1. The temporary memory ANS now contains the value of a_1= -.5. Store this in a memory, say M.

Store the value of 2 as X. Calculate M + the function. The function will be evaluated at 2 and added to the value currently in memory M. The temporary memory ANS now contains a_1 + a_2 = .833. Store this in memory M.

Store the value of 3 as X. Calculate M + the function. The function will be evaluated at 3 and added to the value currently in memory M. The temporary memory ANS now contains a_1 + a_2 + a_3 = 6.083. Store this in memory M.

Continue in this fashion until the memory contains $a_1 + a_2 + a_3 + a_4 + a_5$.
The result is 36.45.

Another method would be to use a separate memory location for the value of the function for each value of x (say memories A, B, C, D, and E). The value of the series could then be calculated as A+B+C+D+E.

Example 2 Textbook Section 10-4

Given the geometric series with a_1 = 2 and r = .85.

(A) Evaluate S_{15} and S_{50} to five decimal place accuracy.

(B) Compare to S_∞. What is the error in each case?

Solution: (A) $S_n = \frac{a_1 - a_1 r^n}{1-r}$. Use A as a_1, R as r, and N as n. Store S_n as an expression in the calculator as (A - A*R ^ N) ÷ (1 − R). Store 2 as A, .85 as R and 15 as N. Evaluate the stored expression. The result is 12.16861. Store this result in a memory, say B, to use in Part (B) below.

Store 50 as N and evaluate the expression again. The result is 13.32939. Store this result in a memory, say C, to use in Part (B) below.

(B) Store S_∞ in the calculator as another expression as: A ÷ (1 − R). Evaluate this expression for the stored values of A and R. The result is 13.33333. Store this result in a memory, say D, to use in finding the error.

Calculate D − B which is the error $S_\infty - S_{15} = 1.16472$.

Calculate D − C which is the error $S_\infty - S_{50} = .00394$.

The second error is smaller because n is larger.

Example 3 Textbook Section 10-5

Evaluate $\binom{32}{14}$.

Solution: $\binom{n}{r} = \frac{n!}{r!(n-r)!}$. One method of solution would be to store the right side of this expression as N ! ÷ (R ! (N − R) !) . Then store 32 as N and 14 as R and evaluate the expression. The result is 471435600.

Another method would be to calculate by entering the numbers into the calculator directly as: 32 ! ÷ (14 ! (32 − 14) !). The result is 471435600. (See Example 1 of Appendix Section A-5, B-5, or C-5 of this manual.)

A third method is to recognize that $\binom{n}{r} = C_{n,r}$ and use the built-in function of the calculator [nCr]. This is found in a menu. It is not a key stroke. Enter 32 [nCr] 14 in the calculator. The result is 471435600. (See Appendix Section A-15, B-15, or C-15 of this manual.)

Example 4 Textbook Section 10-5

Evaluate to two decimal places $(x + y)^{12}$ for $x = 3.58$ and $y = -1.94$.

Solution: Store the expression in the calculator as an expression as (X + Y) ^ 12. Store 3.58 as X and -1.94 as Y. Evaluate the expression. The result is 378.55.

Example 5 Textbook Section 10-5

Evaluate the sixteenth term of $(2x + 3yz^2)^{20}$ for $x = 3.58$, $y = {}^-1.94$, and $z = 5.02$. Use six significant digits.

Solution: The sixteenth term of this expression is $\binom{20}{15}(2x)^5(3yz^2)^{15}$. This can be evaluated directly or by storing the expression in the calculator and then evaluating it.

Evaluating the expression directly

Enter: (20!÷(15!(20-15)!))(2×3.58)^5(3× ⁻1.94× 5.02^2)^15 or
(20 [nCr] 15)(2×3.58)^5(3× ⁻1.94 × 5.02^2)^15 . The result is -9.11938×10^{40} .

Storing the expression

Another method to solve this problem is to store the expression as
(N ! ÷ (R ! (N - R) !)) (A ^ 5)(B ^ 15) . Store 20 as N, 15 as R, 3.58 as X, -1.94 as Y, 5.02 as Z, 2X as A, and 3YZ^2 as B. Evaluate. The result is -9.11938×10^{40} .

An alternate way to store the values for A and B are: Store 2 × 3.58 as A, 3 × ⁻1.94 × 5.02^2 as B. Then evaluate the expression. The result is -9.11938×10^{40}.

Using [nCr]

A third method is to use the built-in function [nCr] . Store (20 [nCr] 15)(A ^ 5)(B ^ 15) . Then store 20 as N, 15 as R, 3.58 as X, -1.94 as Y, 5.02 as Z, 2X as A, and 3YZ^2 as B. Evaluate. The result is -9.11938×10^{40}.

▲▲▲ EXERCISE 10

1. Evaluate the series to two decimal places in Problems 39-44 of textbook Exercise 10-1 using $x = -1.83$.
2. Solve Problems 55 -56 of textbook Exercise 10-1.
3. Solve Problems 1-10 of textbook Exercise 10-2.
4. Verify the series in Problems 23-28 for $n = 5$ of textbook Exercise 10-2.
5. Solve Problems 21-24 of textbook Exercise 10-3.
6. Solve Problems 15-18 and 21-26 of textbook Exercise 10-4.
7. Repeat Example 2 above for $a_1 = 2$ and $r = .89$.

8. Solve Problems 1-12 and 17-24 of textbook Exercise 10-5.

9. Evaluate to nine significant digits the term indicated in Problems 37-44 of textbook Exercise 10-5 using $^{-}5.21$ as the first variable and 1.38 as the second variable in each expression.

NOTES

CHAPTER 11

ADDITIONAL TOPICS IN ANALYTIC GEOMETRY

Example 1 Textbook Section 11-1

Graph $y^2 = -12.5x$.

Solution: To graph $y^2 = -12.5x$ using the calculator it is necessary to solve for y. This results in two functions: $y = \sqrt{-12.5x}$ and $y = -\sqrt{-12.5x}$ for $x \le 0$. Store these functions in the calculator as √ (-12.5 X) and - √ (-12.5 X) . Graph both functions on the same coordinate axes using [-30,30]1 by [-20,20]2. This range was chosen so that the scale marks would be the same distance apart on both axes.

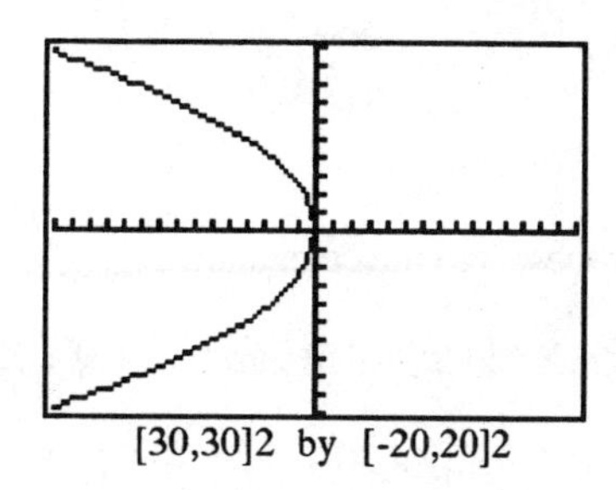

[30,30]2 by [-20,20]2

Example 2 Textbook Section 11-2

Graph $2.4x^2 + 8.5y^2 = 4$. Find the x and y intercepts to three decimal place accuracy.

Solution: It is necessary to solve the equation for y in order to graph using the calculator . This results in two functions:

$y = \sqrt{\dfrac{4 - 2.4x^2}{8.5}}$ and $y = -\sqrt{\dfrac{4 - 2.4x^2}{8.5}}$. Enter these functions into the calculator as √((4-2.4X^2)÷8.5) and ⁻√((4-2.4X^2)÷8.5). Graph both functions on the same coordinate axes using [-3,3]1 by [-2,2]2. This range was chosen so that the scale marks would be the same distance apart on both axes. This gives a more accurate looking graph.

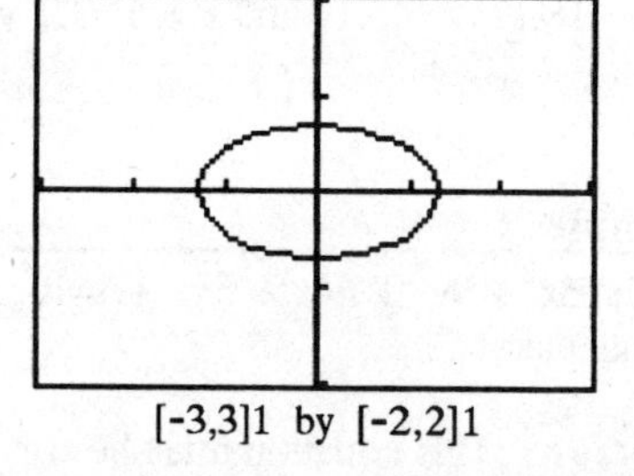

[-3,3]1 by [-2,2]1

Use trace and zoom to find the x and y intercepts. They are (±1.291, 0) and (0, ±.686), respectively.

Example 3 Textbook Section 11-3

Solve the system of equations. Find all coordinates to two decimal places.

$$\frac{x^2}{25} - \frac{y^2}{196} = 13$$

$$30x - 20y^2 = 15$$

Solution: It is necessary to solve both equations for y in order to graph using the calculator. This results in four functions:

$$y = 14\sqrt{\frac{x^2}{25} - 13} \quad \text{and} \quad y = -14\sqrt{\frac{x^2}{25} - 13}$$

$$y = \sqrt{\frac{30x-15}{20}} \quad \text{and} \quad y = -\sqrt{\frac{30x-15}{20}}$$

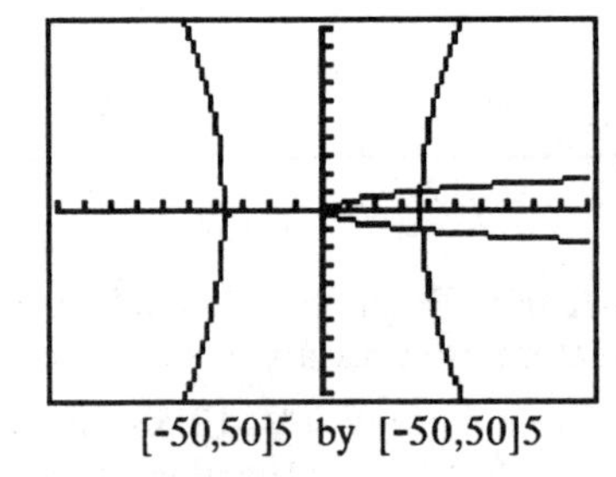

[-50,50]5 by [-50,50]5

Enter these into the calculator as:

14 √ (X^2 ÷ 25 - 13) and -14 √(X^2 ÷ 25 - 13)

√((30 X - 15) ÷ 20) and - √((30 X - 15) ÷ 20)

and graph on the same coordinate axes using [-50,50]5 by [-50,50]5 as shown above.

There are two intersection points and hence two real solutions to this system of equations. Because of symmetry, it is only necessary to find the coordinates of one intersection point. Use trace and zoom to find these coordinates. The result is: (18.12, 5.14). We can find the other intersection point using the symmetry of the graph. It is (18.12,-5.14). Hence, the solutions are $x = 18.12$, $y = 5.14$ and $x = 18.12$, $y = -5.14$.

Example 4 Textbook Section 11-4

Graph $16x^2 + 9y^2 + 64x + 54y + 1 = 0$. Find the x intercepts and y intercepts accurate to three decimal places.

Solution: This equation must be solved for y in order to graph it using the calculator. We must use the quadratic formula to do this. First write the equation with the y^2, y , and terms not involving y clearly identified as:

$$9y^2 + 54y + (16x^2+64x+1) = 0.$$

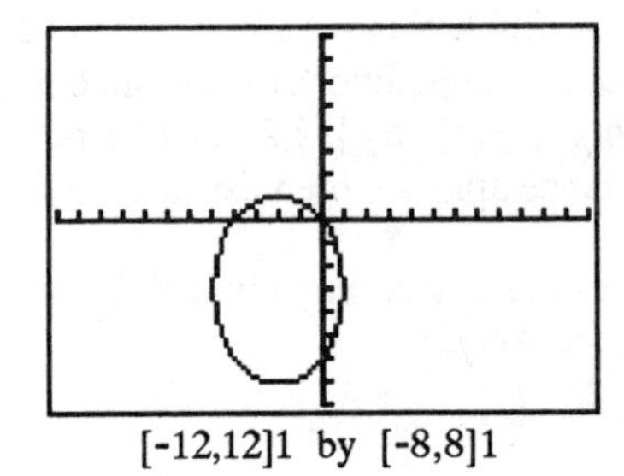

[-12,12]1 by [-8,8]1

Now use the quadratic formula to solve for y:

$$y = \frac{-54 \pm \sqrt{54^2 - 4(9)(16x^2+64x+1)}}{2(9)}.$$

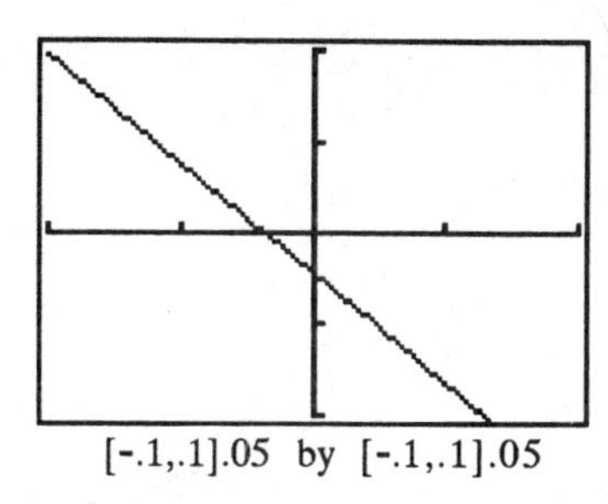
[-.1,.1].05 by [-.1,.1].05

Enter this into the calculator as two separate functions:

(-54+√(54^2-4*9*(16X^2+64X+1))) ÷ (2*9)

and

(-54-√(54^2-4*9*(16X^2+64X+1)))÷ (2*9) .

Graph using [-12,12]1 by [-8,8]1 to get the graph on the preceding page.

Use trace and zoom to find the two x intercepts and two y intercepts. The graph appears to go through the origin. However, changing the RANGE to [-.1,.1].05 by [-.1,.1].05 will show that the graph does not go through the origin.

The graph has x intercepts at (-.016, 0) and (-3.984, 0) and y intercepts at (0, -5.981) and (0, -.019).

▲▲▲ EXERCISE 11

1. Graph Problems 1-12 of textbook Exercise 11-1.
2. Graph the equations in Problems 1-12 of textbook Exercise 11-2.
3. Solve Problems 21-24 of textbook Exercise 11-2.
4. Graph the equations in Problems 1-12 of textbook Exercise 11-3.
5. Solve Problems 21-24 of textbook Exercise 11-3.
6. Graph the equations in Problems 9-14 of textbook Exercise 11-4.
7. Graph the equations in Problems 15-22 of textbook Exercise 11-4. Find the x intercepts and y intercepts of each curve.

NOTES

CHAPTER 12

AN INTRODUCTION TO PROBABILITY

Example 1 Textbook Section 12-1

A company has five departments. Two of these departments has six people and the other three departments have nine people each. In how many ways can a representative and an alternate be chosen from each department to form a committee of five people, each person having an alternate?

Solution: This problem involves both the multiplication principle and permutations. There are $P_{6,2}$ ways to choose two people from a department having six people and $P_{9,2}$ ways to choose two people from a department having nine people. Hence there are:
$(P_{6,2})(P_{6,2})(P_{9,2})(P_{9,2})(P_{9,2})$ ways to choose a committee of five people each having an alternate.

One method to calculate this is to evaluate $P_{6,2}$ using the built-in function of the calculator. (See Appendix Sections A-14, B-14 or C-14 of this manual.) Enter 6 [nPr] 2. Store the result in a memory, say A. Now calculate $P_{9,2}$ by entering 9 [nPr] 2. The result will be in the temporary memory ANS. Now calculate A×A×[ANS]×[ANS]×[ANS]. The result is 335923200. Hence there are 335,923,200 ways to form this committee of five people, one from each department, each person having an alternate.

Another method would be to calculate the result directly using the built-in functions of the calculator. Enter the following into the calculator:
(6 [nPr] 2) (6 [nPr] 2) (9 [nPr] 2)(9 [nPr] 2)(9 [nPr] 2). The result is 335923200.

Example 2 Textbook Section 12-2

There are 30 library books to be sent to three different libraries. The first library is to receive 12 of the books; the second library is to receive 11 of the books; and the third library is to receive the rest. The books are to be chosen in a random manner but there are only 10 ways in which the books could be sent that would satisfy the requirements of each library. What is the probability that the libraries will have their requirements satisfied with the delivery of the books?

Solution: The first step is to identify the sample space and the number in the sample space.

S = {all ways the books can be randomly delivered to the libraries with the first library receiving 12 books, the second library receiving 11 books, and the third library receiving 7 books}

$n(S) = (C_{30,12})(C_{18,11})(C_{7,7})$ [Note: After 12 books are chosen for the first library there are only 18 books left from which to choose the 11 for the second library and then only 7 left from which to choose 7 for the third library.]

The second step is to identify the event and the number in the event.

E = {all ways the books can be delivered that will satisfy the requirements of each library}

$n(E) = 10$

Finally we can calculate the probability: $P(E) = \frac{n(E)}{n(S)} = \frac{10}{(C_{30,12})(C_{18,11})(C_{7,7})}$.

Enter this into the calculator as: 10 ÷ ((30 [nCr] 12) (18 [nCr] 11)). Note that $C_{7,7} = 1$.

The built-in function [nCr] is found as a menu item. The result is $3.632981143 \times 10^{-12}$.

In other words, there is practically no chance that a random selection of books will satisfy the requirements of the libraries.

Example 3 Textbook Section 12-2

A company assigns serial numbers to an item using three letters and five digits. Letters O, I, S and B are not used since they look too much like the digits 0, 1, 5, and 8. What is the probability, to four significant digits, of obtaining a part with a letter prefix of MRQ and digits forming a number less than 500 when choosing a part from those having first letter of M?

Solution: The number in the sample space is $(1)(22^2)(10^5)$ since there is one way to choose the letter M, 22 ways to choose the second letter, 22 ways to choose the third letter, and 10 ways to choose each of the five digits.

The number in the event is (1)(1)(1)(1)(1)(1)(10)(10) since there is one way to choose the M, one way to choose the R, one way to choose the Q, one way to get the ten thousands digit (it has to be a 0); one way to choose the thousands digit (it also has to be a 0); five ways to choose the hundreds digit since numbers must be less than 500 (it can be 0, 1, 2, 3 or 4); 10 ways to choose the tens digit; and 10 ways to choose the units digit.
Hence the probability is:

$$\text{P(E)} = \frac{\text{P(E)}}{\text{P(S)}} = \frac{(1)(1)(1)(1)(1)(10^2)}{(1)(22^2)(10^5)}$$

This can be calculated at one time by entering:

(10 ^ 2) ÷ ((22 ^ 2) (10 ^ 5))

The displayed result on the calculator is 2.06611570248E-06. The desired answer to this problem is then .000002066 accurate to four significant digits.

▲▲▲ EXERCISE 12

1. Solve Problems 1-44 of textbook Exercise 12-1.

2. Solve Problems 51-58 of textbook Exercise 12-2.

3. A library has 45 paintings available for lending. Three patrons request 10, 5 and 8 paintings respectively. If the paintings are to be chosen randomly, how many different possible ways can the requests be filled?

4. A council of 8 churches wishes to form a committee of two people from each of the church's governing boards. In how many ways can this committee be formed if one church has 15 members on the board, another church has 12 members on the board, and the remaining churches each have 9 members on the board?

5. Three dice are to be tossed. What is the probability of getting all dice having the same number?

6. Ten people form a team. A captain, alternate captain, and recorder are chosen from this team. If Paul is the captain, and Carole is the alternate captain, what is the probability that Neal will be chosen as the recorder?

NOTES

APPENDIX A
TI-81 GRAPHING CALCULATOR
BASIC OPERATIONS

A-1 Getting Started

```
RESET
1: No
2: Reset

STAT Bytes        0
PRGM Bytes        0
Bytes Avail    2400
```

Press [ON] to turn on the calculator.

Press [2nd] [+] to get the RESET screen (shown at the right).

Use the down arrow [▼] to choose [2] :Reset and press [ENTER].

The display shows the message **Mem cleared.**

Press [CLEAR] to clear the screen.

Press [2nd] [▲] to make the display darker.

Press [2nd] [▼] to make the display lighter.

Press [2nd] [OFF] to turn off the calculator.

A-2 Calculator Operation

Home Screen

The blank screen is called the Home Screen. You can always get to this screen (aborting any calculations in progress) by pressing [2nd] [QUIT].

[2nd]

This key must be pressed to access the operation above and to the left of a key. These operations are a light blue color on the face of the calculator. An up arrow [↑] is displayed as the cursor on the screen after [2nd] key is pressed. In this manual, the functions above a key will be referred to in square boxes just as if the function was printed on the key cap. For example, [ANS] is the function above the [(-)] key.

ALPHA and A-LOCK

This key must be pressed first to access the operation above and to the right of a key. These operations are in a light grey color on the calculator face. An A is displayed as the cursor on the screen after the ALPHA key is pressed. A-LOCK locks the calculator into alpha mode. The calculator will remain in alpha mode until the A-LOCK is pressed again.

MODE

Press MODE. The highlighted items are active. Select the item you wish using the arrow keys. Press ENTER to activate the selection.

Type of notation for display of numbers.	Norm Sci Eng
Number of decimal places displayed.	Float 0123456789
Type of angle measure.	Rad Deg
Function or parametric graphing.	Function Param
Connected/not connected plotted points on graphs.	Connected Dot
Graph functions separately or all at once.	Sequence Simul
Graph grid display on/off.	Grid Off Grid On
Rectangular or polar coordinate graph.	Rect Polar

Menus

The TI-81 Graphics calculator uses menus for selection of specific functions. The items on the menus are identified by numbers followed by a colon. There are two ways to choose menu items:

1. Using the arrow keys to highlight the selection and then pressing ENTER.
2. Pressing the number corresponding to the menu item.

In this manual the menu items will be referred to using the key to be pressed followed by the meaning of the menu. For example, 1 :Box refers to the first menu item on the ZOOM menu.

A-3 Correcting Errors

It is easy to correct errors when entering data into the calculator by using the arrow keys, INS, and DEL keys.

◀ or ▶	Moves the cursor to the left or right one position.
▲	Moves the cursor up one line or replays the last executed input.
▼	Moves the cursor down one line.
DEL	Deletes one character at the cursor position.
INS	Inserts one character at the cursor position.

A-4 Calculation

Example 1 Calculate $-8 + 9^2 - \left| \frac{3}{\sqrt{2}} - 5 \right|$.

Numbers and characters are entered in the same order as you would read an expression. Do not press [ENTER] unless specifically instructed to do so in these examples. Keystrokes are written in a column but you should enter all the keystrokes without pressing the ENTER key until [ENTER] is displayed in the example.

Solution:

Keystrokes	*Screen Display*	*Explanation*
[2nd] [QUIT]		It is a good idea to clear the screen before starting a calculation.
[(-)] [8] [+] [9] [^] [2] [-]	`⁻8+9^2-abs (3/√2`	
[2nd] [ABS] [(] [3] [÷]	`-5)`	
[2nd] [√] [2] [-] [5] [)]	`70.12132034`	
[ENTER]		

A-5 Evaluation of an Algebraic Expression

Example 1 Evaluate $\frac{x^4 - 3a}{8w}$ for $x = \pi$, $a = \sqrt{3}$, and $w = 4!$.

Two different methods can be used:

1. Store the values of the variables and then enter the expression. When [ENTER] is pressed the expression is evaluated for the stored values of the variables.
2. Store the expression and store the values of the variables.
   Recall the expression. Press [ENTER]. The expression is evaluated for the stored values of the variables.

The advantage of the second method is that the expression can be easily evaluated for several different values of the variables.

Solution:

Keystrokes Method 1	*Screen Display*
[2nd] [QUIT]	
[2nd] [π] [STO▸] [X|T] [ENTER]	`π→X`
	`3.141592654`

> Note that [STO▸] puts the calculator in ALPHA mode automatically.

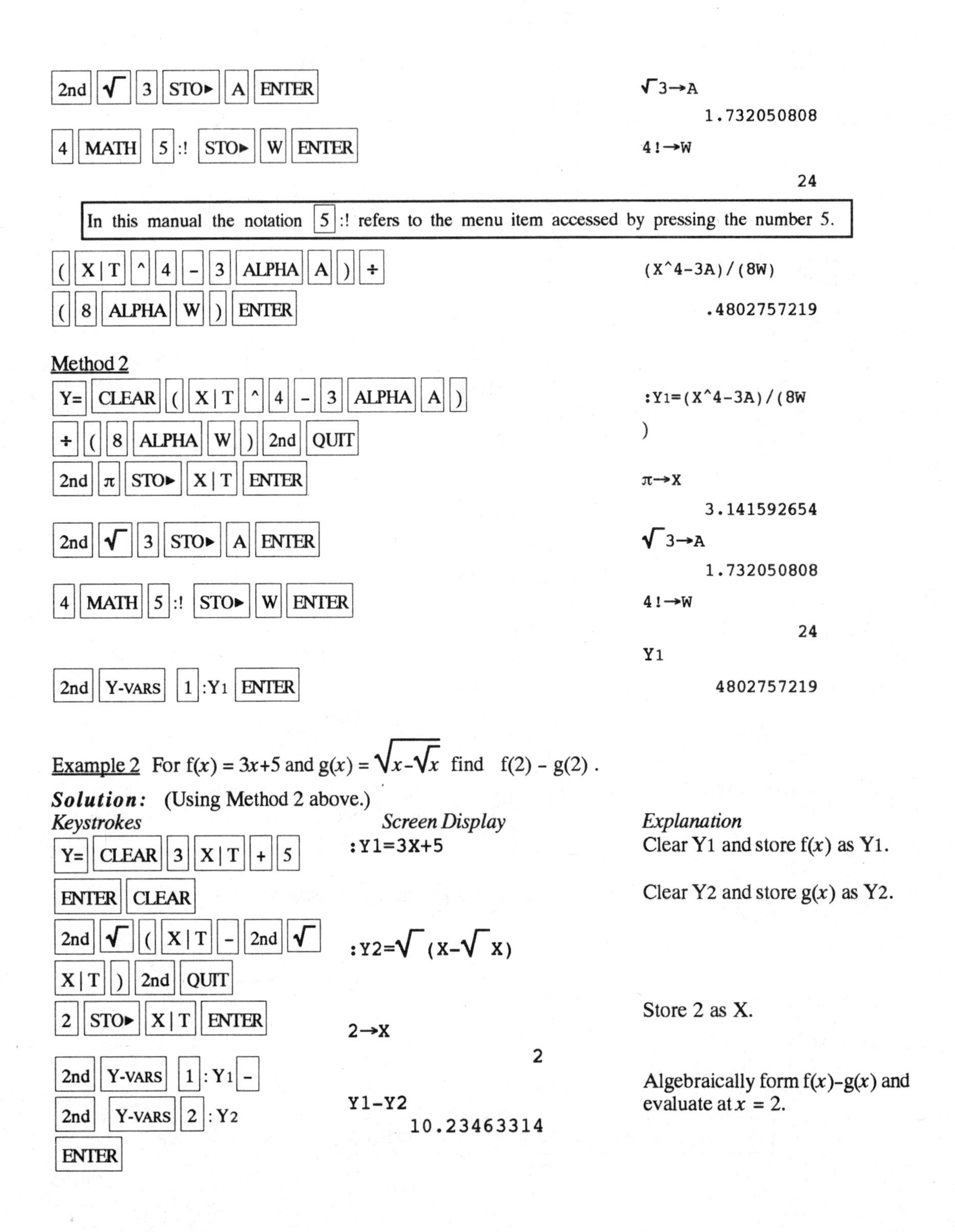

Keystrokes	Screen Display
[2nd] [√] [3] [STO▸] [A] [ENTER]	√3→A 1.732050808
[4] [MATH] [5]:! [STO▸] [W] [ENTER]	4!→W 24

In this manual the notation [5]:! refers to the menu item accessed by pressing the number 5.

Keystrokes	Screen Display
[(] [X|T] [^] [4] [-] [3] [ALPHA] [A] [)] [÷] [(] [8] [ALPHA] [W] [)] [ENTER]	(X^4-3A)/(8W) .4802757219

Method 2

Keystrokes	Screen Display
[Y=] [CLEAR] [(] [X|T] [^] [4] [-] [3] [ALPHA] [A] [)] [÷] [(] [8] [ALPHA] [W] [)] [2nd] [QUIT]	:Y1=(X^4-3A)/(8W)
[2nd] [π] [STO▸] [X|T] [ENTER]	π→X 3.141592654
[2nd] [√] [3] [STO▸] [A] [ENTER]	√3→A 1.732050808
[4] [MATH] [5]:! [STO▸] [W] [ENTER]	4!→W 24
[2nd] [Y-VARS] [1]:Y1 [ENTER]	Y1 4802757219

Example 2 For $f(x) = 3x+5$ and $g(x) = \sqrt{x-\sqrt{x}}$ find $f(2) - g(2)$.

Solution: (Using Method 2 above.)

Keystrokes	*Screen Display*	*Explanation*
[Y=] [CLEAR] [3] [X|T] [+] [5]	:Y1=3X+5	Clear Y1 and store f(x) as Y1.
[ENTER] [CLEAR]		Clear Y2 and store g(x) as Y2.
[2nd] [√] [(] [X|T] [-] [2nd] [√] [X|T] [)] [2nd] [QUIT]	:Y2=√(X-√X)	
[2] [STO▸] [X|T] [ENTER]	2→X 2	Store 2 as X.
[2nd] [Y-VARS] [1]: Y1 [-] [2nd] [Y-VARS] [2]: Y2 [ENTER]	Y1-Y2 10.23463314	Algebraically form f(x)-g(x) and evaluate at x = 2.

A-6 Testing Inequalities in One Variable

Example 1 Determine whether or not $x^3 + 5 < 3x^4 - x$ is true for $x = -\sqrt{2}$.

Solution:

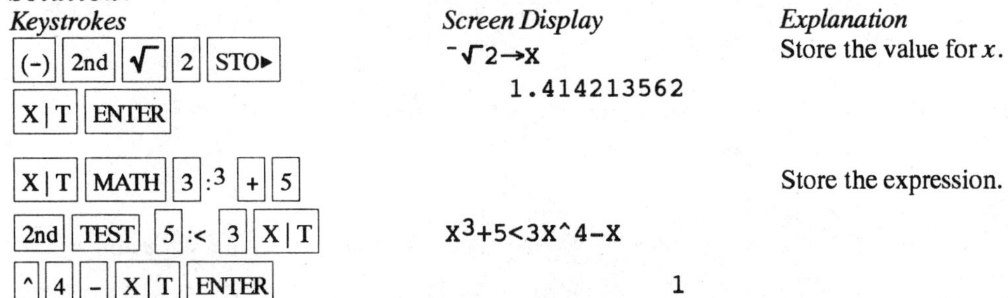

Keystrokes	*Screen Display*	*Explanation*
(-) 2nd √ 2 STO▸	⁻√2→X	Store the value for x.
X|T ENTER	1.414213562	
X|T MATH 3 :³ + 5		Store the expression.
2nd TEST 5 :< 3 X|T	X³+5<3X^4-X	
^ 4 - X|T ENTER	1	

The result of 1 indicates the expression is true for this value of x. If a 0 was displayed, the expression would be false. The expression could have been stored as Y1 and then evaluated as in Method 2 of Example 2 of Section A-5 of this manual.

A-7 Graphing and the Standard Graphing Screen

Up to four functions can be stored and graphed on the same coordinate axes.

Example 1 Graph $y = x^2$, $y = .5x^2$, $y = 2x^2$, and $y = -1.5x^2$ on the same coordinate axes.

Solution:

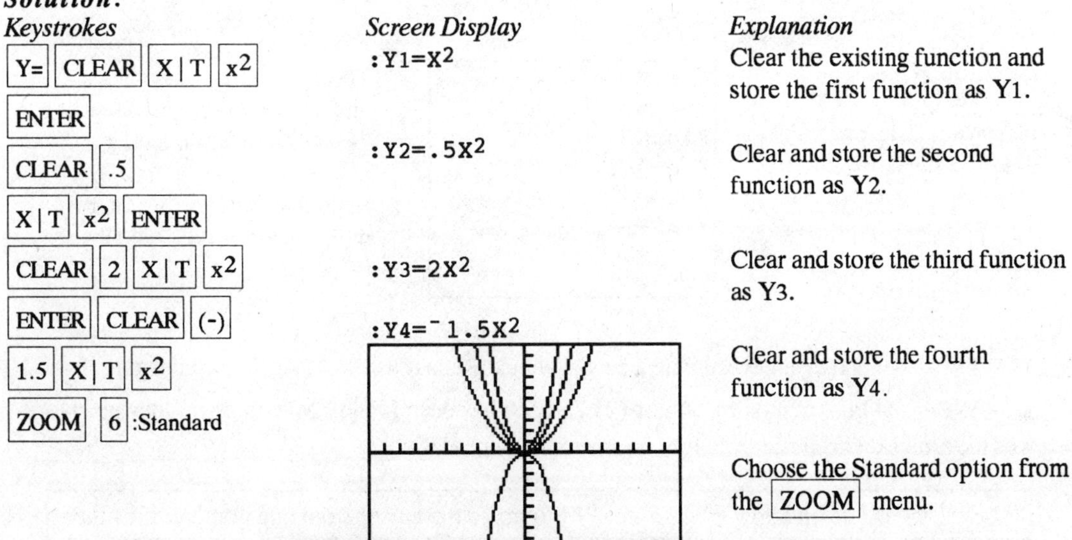

Keystrokes	*Screen Display*	*Explanation*
Y= CLEAR X|T x²	:Y1=X²	Clear the existing function and store the first function as Y1.
ENTER		
CLEAR .5	:Y2=.5X²	Clear and store the second function as Y2.
X|T x² ENTER		
CLEAR 2 X|T x²	:Y3=2X²	Clear and store the third function as Y3.
ENTER CLEAR (-)		
1.5 X|T x²	:Y4=⁻1.5X²	Clear and store the fourth function as Y4.
ZOOM 6 :Standard		Choose the Standard option from the ZOOM menu.

The Standard screen automatically sets the graph for $-10 < x < 10$ and $-10 < y < 10$. Press RANGE to see this.

The graphs will be plotted in order: Y1, then Y2, then Y3, then Y4.

A-8 TRACE, ZOOM and RANGE

TRACE allows you to observe both the x and y coordinate of a point on the graph as the cursor moves along the graph. If there is more than one function graphed the up ▲ and down ▼ arrow keys allow you to move between the graphs displayed.

ZOOM will magnify a graph so the coordinates of a point can be approximated with greater accuracy. There are three methods to zoom in:

1. Change the RANGE values.
2. Use the 2 :Zoom In option on the ZOOM menu in conjunction with 4 :Set Factors.
3. Use the 1 :Box option on the ZOOM menu.

Example 1 Approximate the value of x to two decimal places if $y = -1.58$ for $y = x^3 - 2x^2 + \sqrt{x} - 8$.

Solution:

Graph the function using the Standard Graphing Screen (See Section A-7 of this manual).

Method 1 Change the RANGE values.

Keystrokes	*Screen Display*	*Explanation*
		Press the right arrow repeatedly until the new type of cursor gives a y value as close as possible to -1.58 which is (2.63,-2.00)
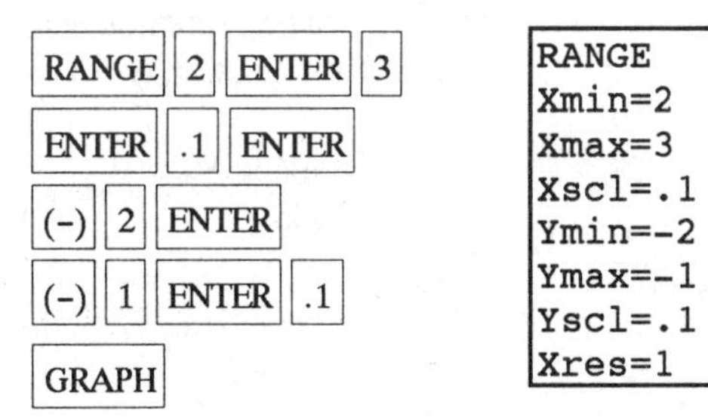	RANGE Xmin=2 Xmax=3 Xscl=.1 Ymin=-2 Ymax=-1 Yscl=.1 Xres=1	The x coordinate is between 2 and 3. So we set the RANGE at $2<x<3$ with scale marks every .1 by $-2<y<-1$ with scale marks every .1. This will be written as [2,3].1 by [-2,-1].1.

TRACE can be used again to estimate a new x value. Repeat using TRACE and changing the RANGE until the approximation of (2.67,-1.58) has been found. Note that you may need to press the arrow keys repeatedly before the cursor becomes visible.

> Use the up and down arrow keys ▲ or ▼ to move the cursor from one graph to the other.

> Occasionally you will see an array of dots in the upper right corner. This means the calculator is working. Wait until the dot array disappears before continuing.

Method 2 Use the [2]:Zoom In option on the [ZOOM] menu.

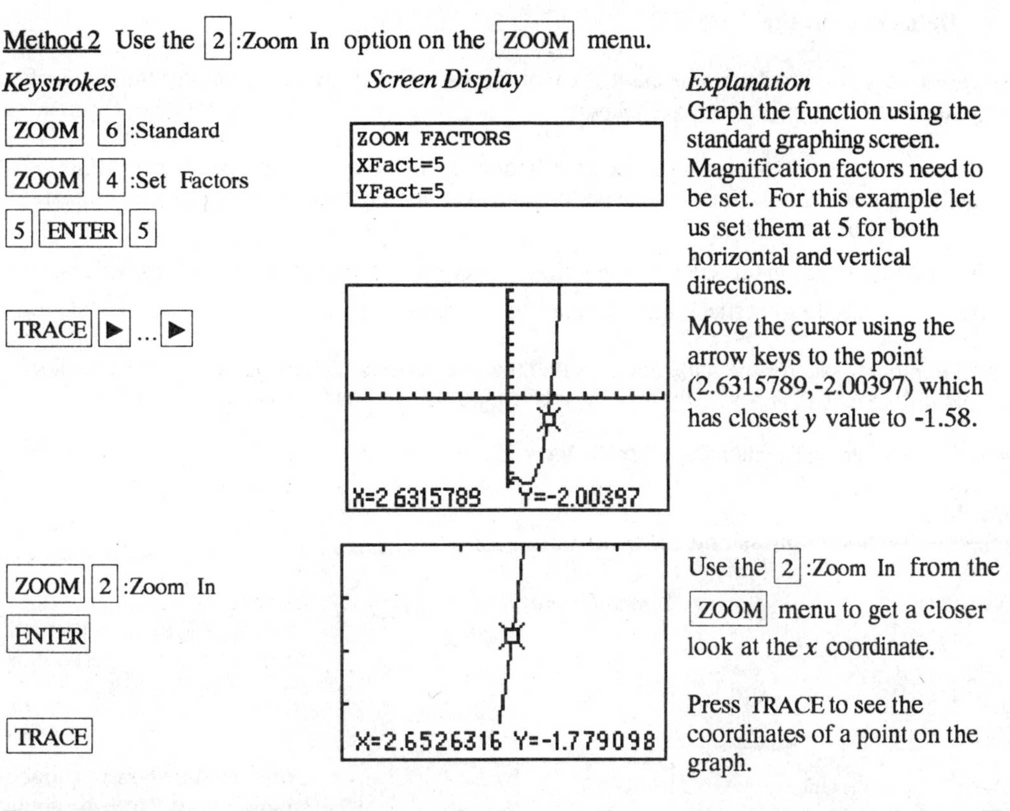

Keystrokes	*Screen Display*	*Explanation*
[ZOOM] [6]:Standard [ZOOM] [4]:Set Factors [5] [ENTER] [5]	ZOOM FACTORS XFact=5 YFact=5	Graph the function using the standard graphing screen. Magnification factors need to be set. For this example let us set them at 5 for both horizontal and vertical directions.
[TRACE] [▶] ... [▶]	X=2.6315789 Y=-2.00397	Move the cursor using the arrow keys to the point (2.6315789,-2.00397) which has closest y value to -1.58.
[ZOOM] [2]:Zoom In [ENTER]		Use the [2]:Zoom In from the [ZOOM] menu to get a closer look at the x coordinate.
[TRACE]	X=2.6526316 Y=-1.779098	Press TRACE to see the coordinates of a point on the graph.

Repeat this procedure until you get a value for the x coordinate accurate to two decimal places. The point has coordinates (2.67,-1.58).

Method 3 Use the Zoom Box option on the [ZOOM] menu.

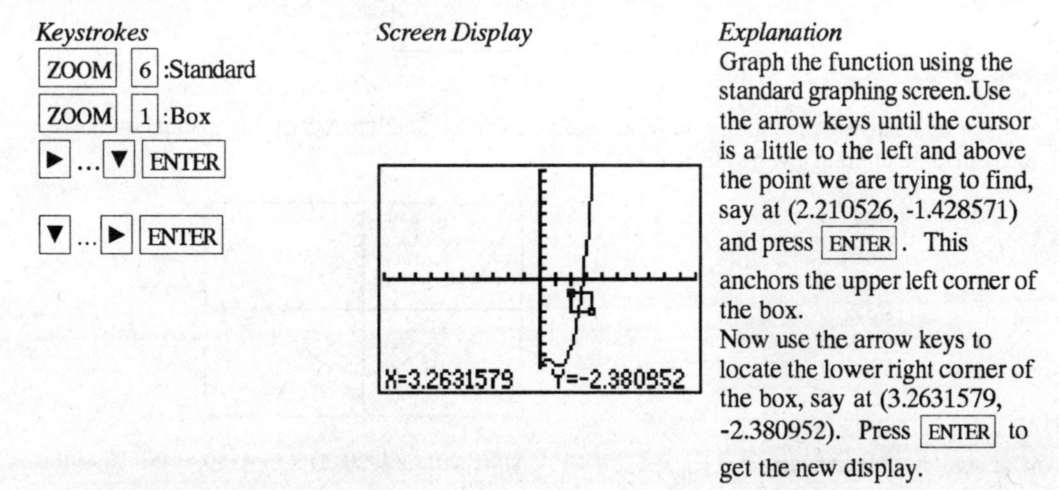

Keystrokes	*Screen Display*	*Explanation*
[ZOOM] [6]:Standard [ZOOM] [1]:Box [▶] ... [▼] [ENTER]		Graph the function using the standard graphing screen. Use the arrow keys until the cursor is a little to the left and above the point we are trying to find, say at (2.210526, -1.428571) and press [ENTER]. This anchors the upper left corner of the box.
[▼] ... [▶] [ENTER]	X=3.2631579 Y=-2.380952	Now use the arrow keys to locate the lower right corner of the box, say at (3.2631579, -2.380952). Press [ENTER] to get the new display.

Repeat this procedure until you get a value for the x coordinate accurate to two decimal places. The point has coordinates (2.67,-1.58).

A-9 Determining the RANGE

There are several ways to determine the RANGE of the values that should be used for the limits of the x and y axes. Three are described below:

1. Graph using the default setting of the calculator and zoom out. The disadvantage of this method is that often the function cannot be seen at either the default settings or the zoomed out settings of the RANGE.
2. Evaluate the function for several values of x. Make a first estimate based on these values.
3. Analyze the leading coefficient and the constant terms.

A good number to use for the scale marks is one that yields about 20 marks across the axis. For example if the RANGE is [-30,30] for the x axis a good scale value is (30–(-30))/20 or 3.

Example 1 Graph the function $f(x)= .2x^2 + \sqrt[3]{x} - 32$.

Solution:

Method 1 Use the default setting and zoom out.

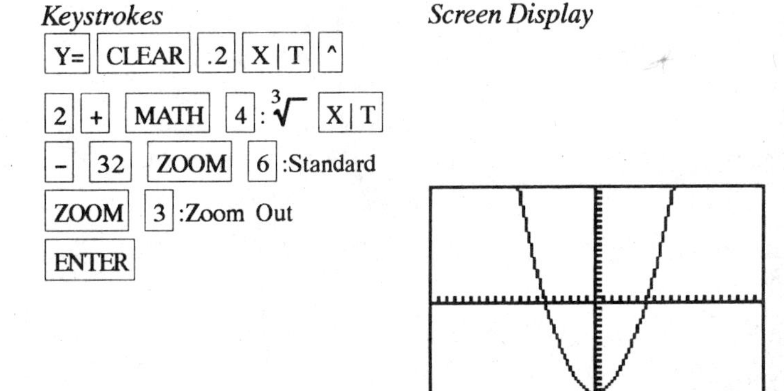

Explanation

Nothing is seen on the graph screen because no part of this curve is in this RANGE.

Zooming out shows a parabolic shaped curve. Note the double axis. This indicates that the scale marks are very close together.

Method 2 Evaluate the function for several values of x. (See Section A-5 of this manual on how to evaluate a function at given values of x.)

x	f(x)
-20	45.3
-10	-14.2
0	-32.0
10	- 9.8
20	50.7

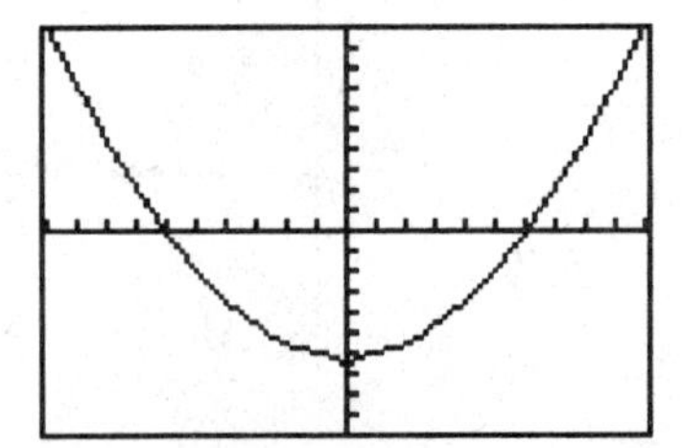

Analyzing this table indicates that a good RANGE to start with is [-20,20]2 by [-50,50]5. Note the scale is chosen so that about 20 scale marks will be displayed along each of the axes.

Method 3 Analyze the leading coefficient and constant terms.
Since the leading coefficient is .2 the first term will increase 2 units for each 10 units x^2 increases ($\sqrt{10}$ or about 3 units increase in x). A first choice for the x-axis limits can be

$$\frac{10\times(\text{unit increase in } x)}{(\text{first term increase})} = \frac{10\times 3}{2} = 15$$

A first choice for the scale on the x axis (having about 20 marks on the axis) can be found using $\frac{\text{Xmax-Xmin}}{20} = \frac{15-(-15)}{20} = 1.5$ (round to 2). So the limits on the x axis could be [-15,15]2.

A first choice for the y-axis limits could be ±(constant term). The scale for the y axis can be found using $\frac{\text{Ymax-Ymin}}{20} = \frac{32-(-32)}{20} = 3.2$ (round to 4). So a first choice for the y-axis limits could be [-32,32]4. Hence a good first setting for the the RANGE if [-15,15]2 by [-32,32]4.

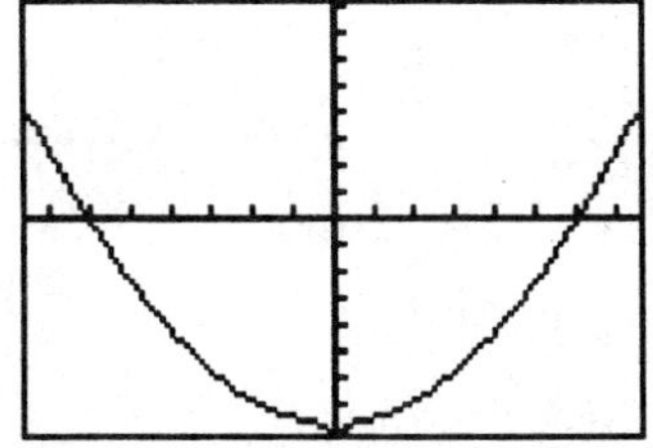

A good choice for the **scale** is so that about 20 marks appear along the axis. This is $\frac{\text{Xmax-Xmin}}{20}$ (rounded up to the next integer) for the x axis and $\frac{\text{Ymax-Ymin}}{20}$ (rounded up to the next integer) for the y axis.

A-10 Piecewise-Defined Functions

There are two methods to graph piecewise-defined functions:

1. Graph each piece of the function separately as an entire function on the same coordinate axes. Use trace and zoom to locate the partition value on each of the graphs.
2. Store each piece of the function separately but include an inequality statement following the expression which will set the RANGE of values on x for which the function should be graphed. Then graph all pieces on the same coordinate axes.

Example 1 Graph $f(x) = \begin{cases} x^2+1 & x < 1 \\ 3x-5 & x \geq 1 \end{cases}$

Solution:

Method 1

Keystrokes	*Screen Display*	*Explanation*
Y= CLEAR X|T ^ 2 + 1 ENTER CLEAR 3 X|T - 5 ZOOM 6 :Standard	:Y1=X^2+1 :Y2=3X-5 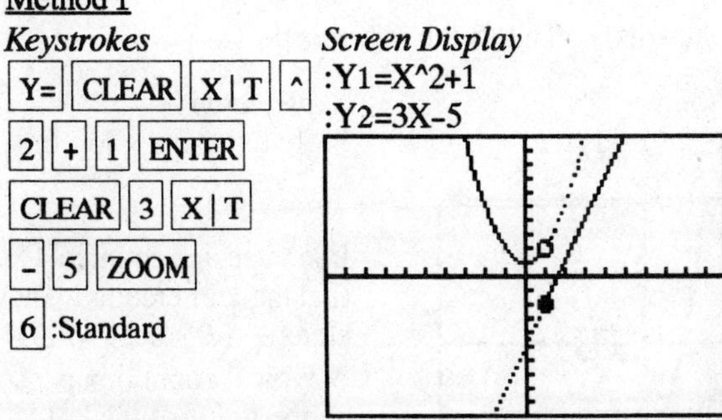	Store the functions. Graph. Both functions will be displayed. Use trace and zoom to find the point on the graphs where x=1. Place an open circle on as the endpoint of the piece of the graph not including x=1 and a closed circle as the endpoint of the piece of the graph including x=1. The graph shown to the left shows the curves with the partition values indicated. This result cannot be displayed on the calculator; it must be added by hand.

Method 2

Keystrokes	*Screen Display*	*Explanation*
Y= CLEAR (X|T ^ 2 + 1) ÷ (X|T 2nd TEST 5 :< 1) ENTER CLEAR (3 X|T - 5) ÷ (X|T 2nd TEST 4 :≥ 1) ZOOM 6 :Standard	:Y1=(X^2+1)/(X<1) :Y2:(3X-5)/(X≥1)	The logical statement $x<1$ will give a 1 when the value of x is less than 1 and a 0 when the value of x is greater than or equal to 1. Hence the first part of the function is divided by 1 when $x<1$ and 0 when $x\geq1$. The function will not graph when it is divided by 0. Similarly for the logical statement $x\geq1$ for the second part of the function. The 1 and 0 are not shown on the screen but are used by the calculator when graphing the functions.

The calculator will display a line from the end of one function to the beginning of the other function. Change the mode to DOT to eliminate this line. Another way to avoid this is to store the function as (X^2+1)(1÷(X<1)). The expression (1÷(X<1)) is undefined when x is less than 1 and hence this piece of the function will not be graphed for x < 1.

A-11 Solving Equations in One Variable

There are two methods for approximating the solution of an equation using graphing.

1. Write the equation as an expression equal to zero. Graph y=(the expression). Find where the curve crosses the x axis. These x values are the solution to the equation.
2. Graph y = (left side of the equation) and y=(right side of the equation) on the same coordinate axes. The x coordinate of the points of intersection are the solutions to the equation.

Example 1 Solve $\frac{3x^2}{2} - 5 = \frac{2(x+3)}{3}$.

Solution:

Method 1 Write the equation as $\left(\frac{3x^2}{2} - 5\right) - \left(\frac{2(x+3)}{3}\right) = 0$. Graph $y = \left(\frac{3x^2}{2} - 5\right) - \left(\frac{2(x+3)}{3}\right)$ and find the x value where the graph crosses the x axis. This is the x intercept.

Keystrokes

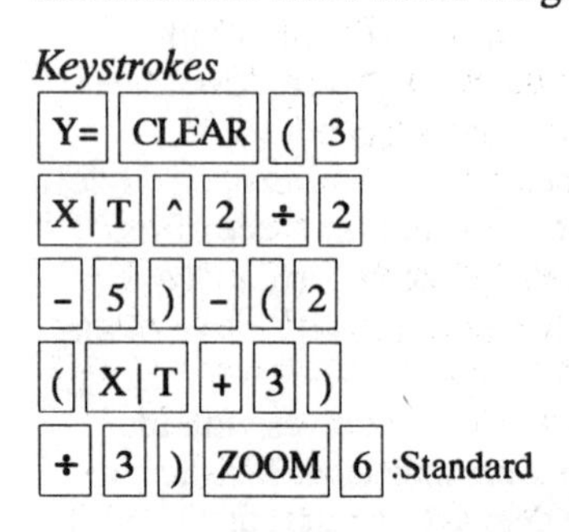

Screen Display

:Y1=(3X^2/2-5)-(
2(X+3)/3)

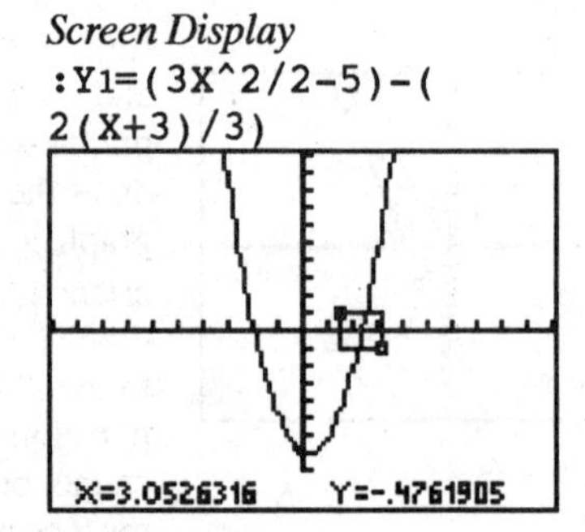

Explanation

Store the expression as Y1.

Use trace and zoom to find the x intercepts. They are: $x \approx -1.95$ and $x \approx 2.39$. A typical zoom box is shown on the graph at the left.

<u>Method 2</u> Graph $y = \frac{3x^2}{2} - 5$ and $y = \frac{2(x+3)}{3}$ on the same coordinate axes and find the x coordinate of their points of intersection.

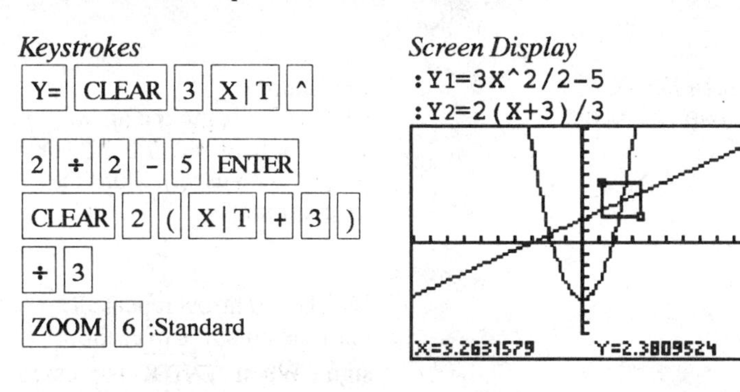

Explanation
Store the two functions.

Find the points of intersection. Use trace and zoom to find the x values: $x \approx -1.95$ and $x \approx 2.39$.

A typical zoom box is shown on the graph at the left.

Hence the approximate solutions to this equation are -1.95 and 2.39.

A-12 Solving Inequalities in One Variable

There are two methods for approximating the solution of an inequality using graphing.

1. Write the inequality with zero on one side of the inequality sign. Graph y=(the expression). Find the x intercepts. The solution will be an inequality with the x values (x intercepts) as the cut off numbers.

2. Graph y=(left side of the inequality) and y=(right side of the inequality) on the same coordinate axes. The x coordinate of the points of intersection are the solutions to the equation. Identify which side of the x value satisfies the inequality by observing the graphs of the two functions.

<u>Example 1</u> Approximate the solution to $\frac{3x^2}{2} - 5 \le \frac{2(x+3)}{3}$. Use two decimal place accuracy.

Solution:

<u>Method 1</u> Write the equation as $\left(\frac{3x^2}{2} - 5\right) - \left(\frac{2(x+3)}{3}\right) \le 0$.

Graph $y = \left(\frac{3x^2}{2} - 5\right) - \left(\frac{2(x+3)}{3}\right)$ and find the x intercepts.

This was done in Example A-6 Method 1 of this manual. The x intercepts are -1.95 and 2.39. The solution to the inequality is the interval on x for which the graph is below the x axis. The solution is $-1.95 \le x \le 2.39$.

<u>Method 2</u> Graph $y = \frac{3x^2}{2} - 5$ and $y = \frac{2(x+3)}{3}$ on the same coordinate axes and find the x coordinate of their points of intersection. This was done in Example A-6 Method 2 of this manual. The x coordinate of the points of intersections are -1.95 and 2.39. We see that the parabola is below the x axis for $-1.95 \le x \le 2.39$. Hence the inequality is satisfied for $-1.95 \le x \le 2.39$.

A-13 Storing an Expression That Will Not Graph

Example 1 Store the expression B^2-4AC so that it will not be graphed but so that it can be evaluated at any time.

Solution:

Keystrokes	*Screen Display*	*Explanation*
[Y=] [▼] [▼] [▼] [4] :Y4 [CLEAR] [ALPHA] [B] [^] [2] [-] [4] [ALPHA] [A] [ALPHA] [C]	:Y4=B^2-4AC	Choose Y4 using the arrow keys. (Any of Y1, Y2, Y3, or Y4 could be used.) Store the expression.
[◀] ... [◀] [ENTER]		Use the left arrow repeatedly until the cursor is over the = sign. When [ENTER] is pressed the highlighting will disappear from the = sign. Now you can still evaluate the expression using [Y-VARS] but it will not graph. (See Section A-5 Example 2 of this manual.)

A-14 Permutations and Combinations

Example 1 Find (A) $P_{10,3}$ and (B) $C_{12,4}$

Solution (A):

Keystrokes

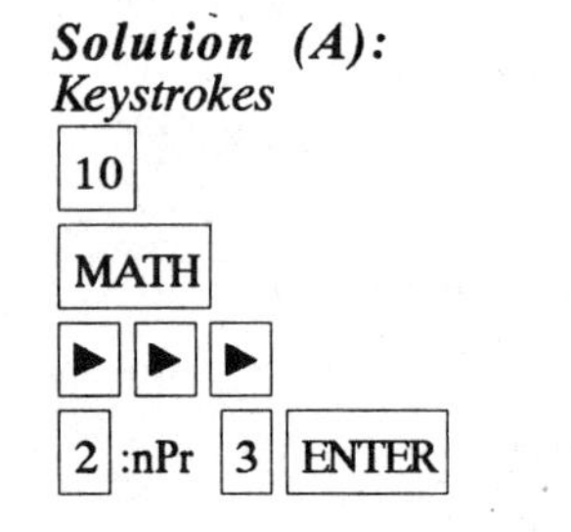

Screen Display

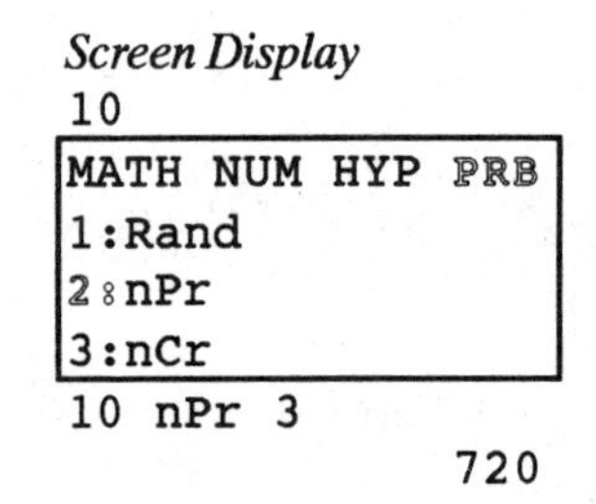

Explanation

Enter the first number. Get the math menu and choose PRB using the arrow keys. Choose nPr and press [ENTER].

Solution (B):

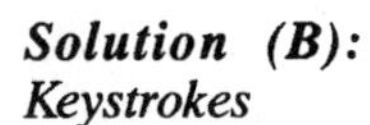

Keystrokes

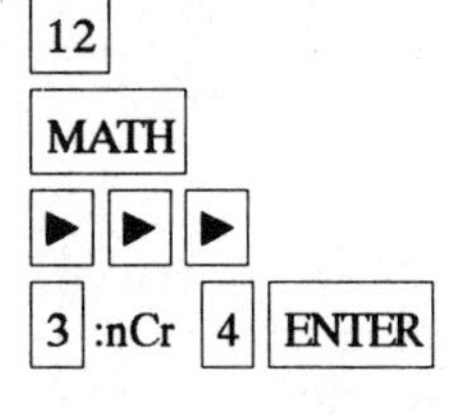

Screen Display

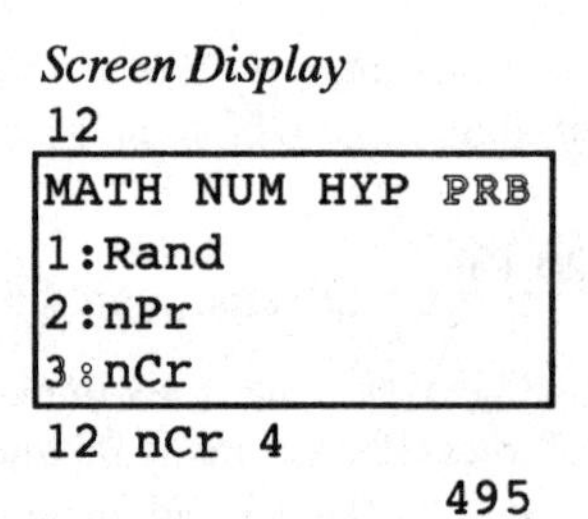

Explanation

Enter the first number. Get the math menu and choose PRB using the arrow keys. Choose nCr and press [ENTER].

A-15 Matrices

Example 1 Given the matrices

$$A = \begin{bmatrix} 1 & -2 \\ 3 & 0 \\ 5 & -8 \end{bmatrix} \quad B = \begin{bmatrix} 2 & 1 & 5 \\ 3 & 2 & -1 \\ 0 & 8 & -3 \end{bmatrix} \quad C = \begin{bmatrix} 1 \\ -5 \\ 10 \end{bmatrix}$$

Find (A) -3BC (B) B^{-1} (C) A^T (D) det B

Solution (A):

Keystrokes	*Screen Display*	*Explanation*
MATRX ▶	MATRIX EDIT 1:RowSwap(2:Row+(3:*Row(4:*Row+(5:det 6:T	Enter the matrix mode. Choose EDIT using the arrow keys.
1 :[A]	MATRIX EDIT 1: [A] 3x3 2: [B] 3x2 3: [C] 3x1	Choose the A matrix.
3 ENTER 2 ENTER		Store the dimensions of the matrix.
1 ENTER (-) 2 ENTER 3 ENTER 0 ENTER 5 ENTER (-) 8 ENTER	[A] 3x2 1,1=1 1,2=-2 2,1=3 2,2=Ø 3,1=5 3,2=-8	Enter the matrix elements. You will notice a dot array at the top right of the screen. This has the same number of dots as the size of your matrix. The blank dot is the location at which you are entering a number.
MATRX		Return to the matrix menu and repeat the procedure to enter matrix B and C.
2nd QUIT 2nd [A] ENTER	[A] [1 -2] [3 Ø] [5 -8]	Return to the home screen. Recall matrix A. On the home screen you will now see the row/column form.

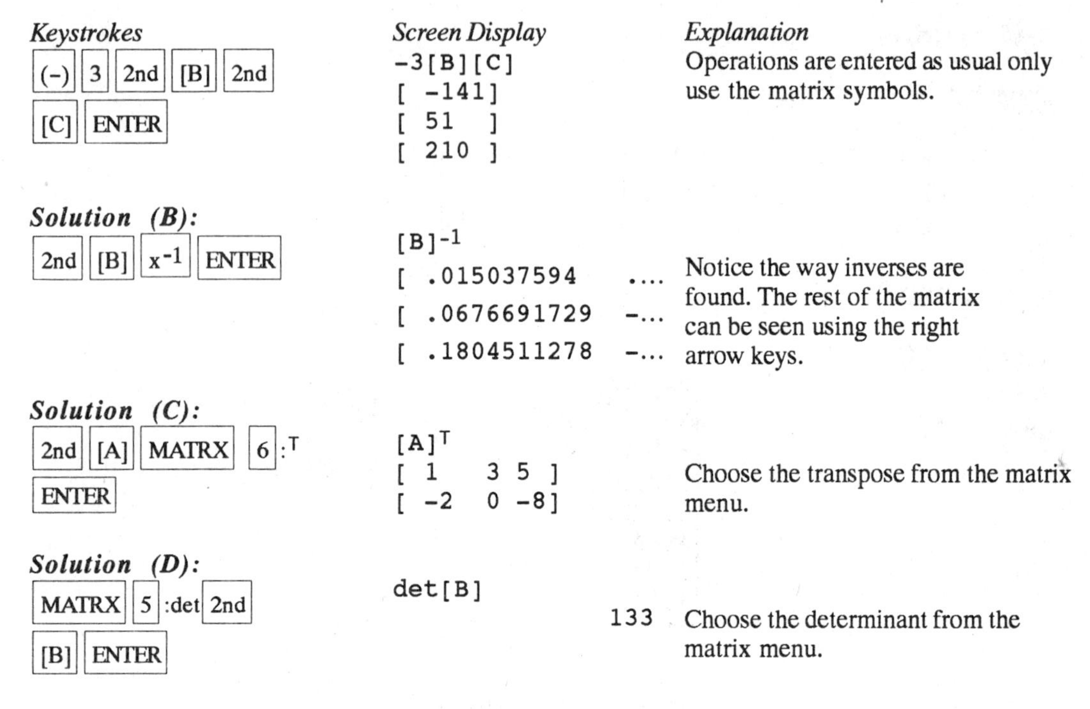

Keystrokes	*Screen Display*	*Explanation*
(-) 3 2nd [B] 2nd [C] ENTER	-3[B][C] [-141] [51] [210]	Operations are entered as usual only use the matrix symbols.

Solution (B):

Keystrokes	*Screen Display*	*Explanation*
2nd [B] x^{-1} ENTER	[B]$^{-1}$ [.015037594 .... [.0676691729 -... [.1804511278 -...	Notice the way inverses are found. The rest of the matrix can be seen using the right arrow keys.

Solution (C):

Keystrokes	*Screen Display*	*Explanation*
2nd [A] MATRX 6 :T ENTER	[A]^T [1 3 5] [-2 0 -8]	Choose the transpose from the matrix menu.

Solution (D):

Keystrokes	*Screen Display*	*Explanation*
MATRX 5 :det 2nd [B] ENTER	det[B] 133	Choose the determinant from the matrix menu.

<u>Example 2</u> Find the reduced form of matrix $\begin{bmatrix} 2 & 1 & 5 & 1 \\ 3 & 2 & -1 & -5 \\ 0 & 8 & -3 & 10 \end{bmatrix}$

Store the matrix dimensions and elements using the procedure in Example 1 of Section A-15. Notice the arrows ↑ and ↓ on the screen indicating there are more elements above and below the display. Use the arrow keys to see the other elements.

Solution:

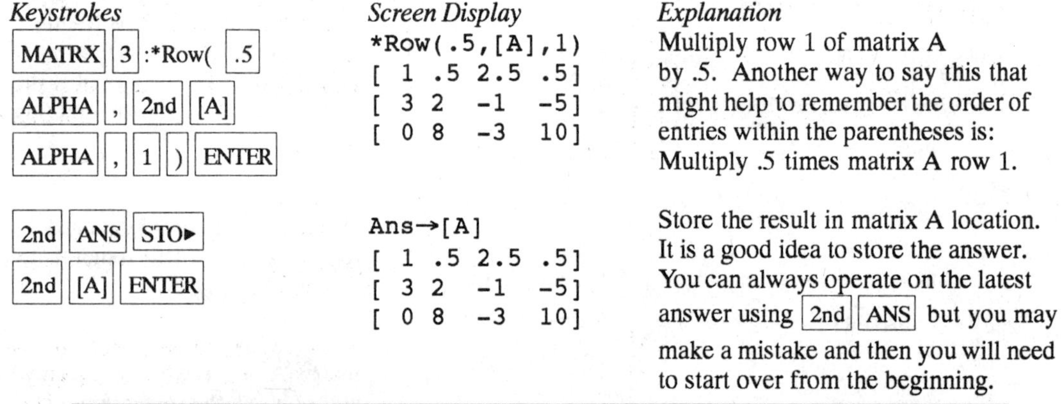

Keystrokes	*Screen Display*	*Explanation*
MATRX 3 :*Row(.5 ALPHA , 2nd [A] ALPHA , 1) ENTER	*Row(.5,[A],1) [1 .5 2.5 .5] [3 2 -1 -5] [0 8 -3 10]	Multiply row 1 of matrix A by .5. Another way to say this that might help to remember the order of entries within the parentheses is: Multiply .5 times matrix A row 1.
2nd ANS STO▸ 2nd [A] ENTER	Ans→[A] [1 .5 2.5 .5] [3 2 -1 -5] [0 8 -3 10]	Store the result in matrix A location. It is a good idea to store the answer. You can always operate on the latest answer using 2nd ANS but you may make a mistake and then you will need to start over from the beginning.

REMINDER: Commas are needed between entries in matrix row operations commands.

MATRX 4 :*Row+(
(-) 3 ALPHA ,
2nd [A] ALPHA , 1
ALPHA , 2) ENTER

```
*Row+(-3,[A],1,2
)
[ 1 .5 2.5    .5 ...]
[ 0 .5 -8.5   -6 ...]
[ 0 8  -3     10 ...]
```

Multiply -3 times matrix A row 1 to add to row 2.

2nd ANS STO▸
2nd [A] ENTER

```
Ans→[A]
[ 1 .5   2.5  .5 ...]
[ 0 .5   -8.5 -6 ...]
[ 0 8    -3   10 ...]
```

Store the result.
Check the result.

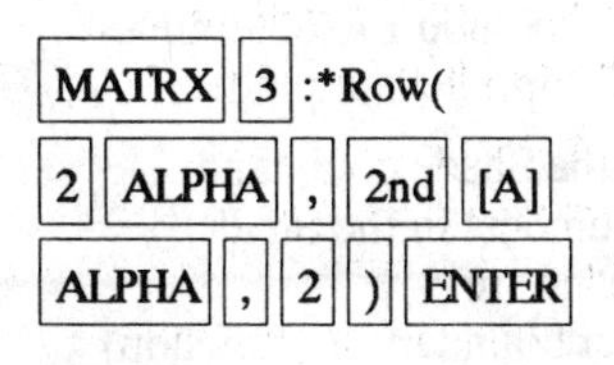


```
*Row(2,[A],2)
[ 1 .5 2.5  .5 ]
[ 0 1  -17  -13]
[ 0 8  -3   10 ]
```

2 times matrix A row 2.

2nd ANS STO▸
2nd [A] ENTER

```
Ans→[A]
[ 1 .5 2.5  .5 ]
[ 0 1  -17  -13]
[ 0 8  -3   10 ]
```

Store the result.
View the result.

Continue using row operations to arrive at the reduced form of $\begin{bmatrix} 1 & 0 & 0 & -2.428... \\ 0 & 1 & 0 & 1.571... \\ 0 & 0 & 1 & .857... \end{bmatrix}$

Note: To swap rows of a matrix use 1 :RowSwap(from the MATRX menu. RowSwap([A],2,3) will swap rows 2 and 3 in matrix A.

A-16 Graphing an Inequality

There are two methods to graph an inequality.

1. Graph the boundary curve. Determine the half-plane by choosing a test point not on the boundary curve and substituting into the inequality.
2. Repeat Method 1 to determine which side of the graph is to be shaded. Use the SHADE option on the calculator to get a shaded graph.

Example 1 Graph $3x + 4y \le 12$

Solution:

Keystrokes

Method 1

[Y=] [(] [12]
[-] [3] [X|T]
[)] [÷] [4]
[ZOOM] [6] :Standard

Screen Display

```
:Y1=(12-3X)/4
```

Explanation

Graph $3x+4y=12$ by first writing as $y=(12-3x)/4$. Determine the half-plane by choosing the point (0,0) and substituting into the inequality by hand. The inequality is true for this point. Hence, we want the lower half-plane.

Method 2

Repeat Method 1 to determine the appropriate half plane.

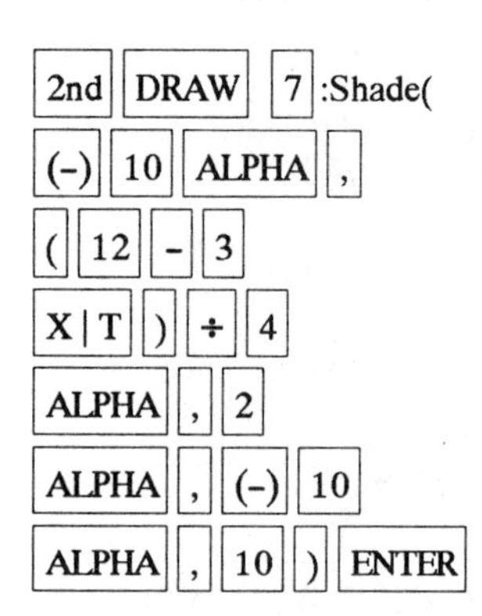


```
Shade(-10,(12-3x
)/4,2,-10,10)
```

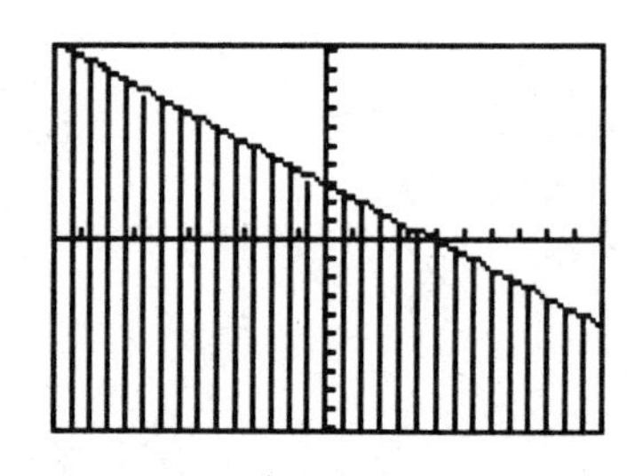

Enter the Shade command. The numbers in the Shade command are

Lower boundary	(a function)
Upper boundary	(a function)
Resolution	(a number from 1 to 8 for shading pattern.)
Left boundary	(a number)
Right boundary	(a number)

REMINDER: Commas are needed between entries in the shade command.

A-17 Exponential and Hyperbolic Functions

Example 1 Graph $y = 10^{0.2x}$

Solution:

Keystrokes

[Y=] [CLEAR] [10] [^] [(] [.2]
[X|T] [)] [ZOOM] [6] :Standard

Screen Display

```
:Y1=10^(.2X)
```

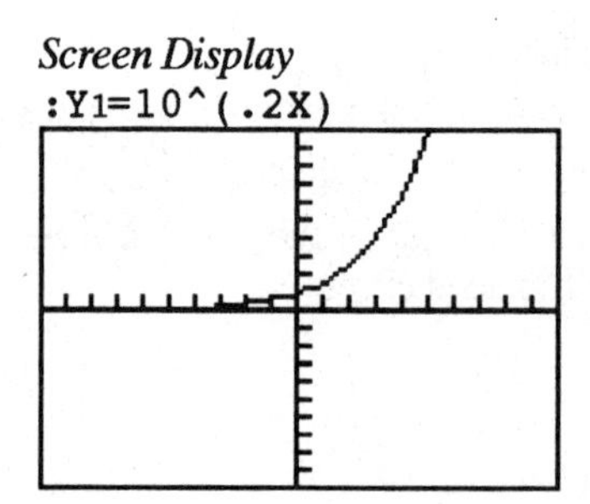

Explanation

Store the function and graph. Note the entire exponent needs to be in parentheses.

Example 2 Graph $y = \frac{e^{x} - e^{-x}}{2}$.

Solution:

Keystrokes

Y= CLEAR MATH ► ►

1 :sinh

X | T ZOOM 6

Screen Display

```
:Y1=sinh X
MATH NUM HYP
PRB
1:sinh
2:cosh
3:tanh
4:sinh-1
5:cosh-1
6:tanh-1
```

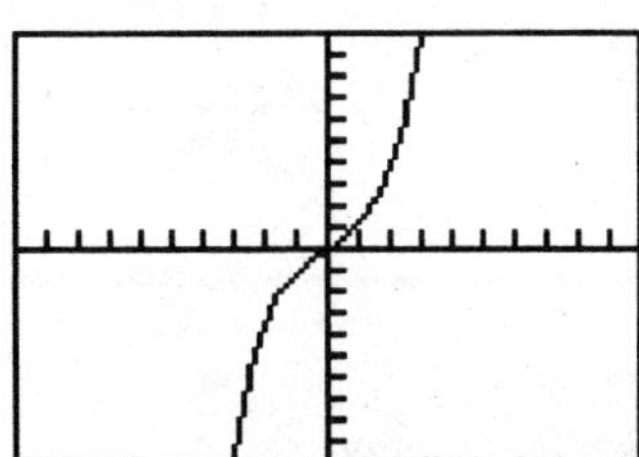

Explanation

We observe that this is the hyperbolic sine function. So we can use the built-in function in the calculator.

The function could also have been graphed by storing:

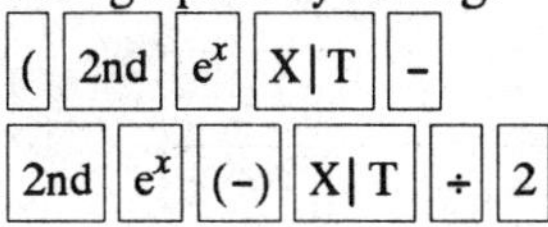

(2nd e^x X | T –

2nd e^x (–) X | T ÷ 2

as a function and graphing.

A-18 Angles and Trigonometric Functions

Example 1 Evaluate $f(x) = \sin x$ and $g(x) = \tan^{-1}x$ at $x = \frac{5\pi}{8}$.

Solution:

Keystrokes

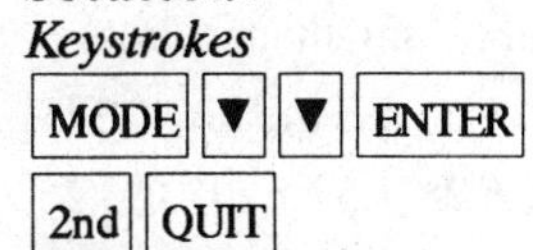

MODE ▼ ▼ ENTER

2nd QUIT

Screen Display

```
Norm Sci Eng
Float 0123456789
Rad Deg
Function Param
Connected Dot
Sequence Simul
Grid Off Grid On
Rect Polar
```

Explanation

The angle measure is given in radians. Set the calculator for radian measure before starting calculations. Return to the Home screen using 2nd QUIT.

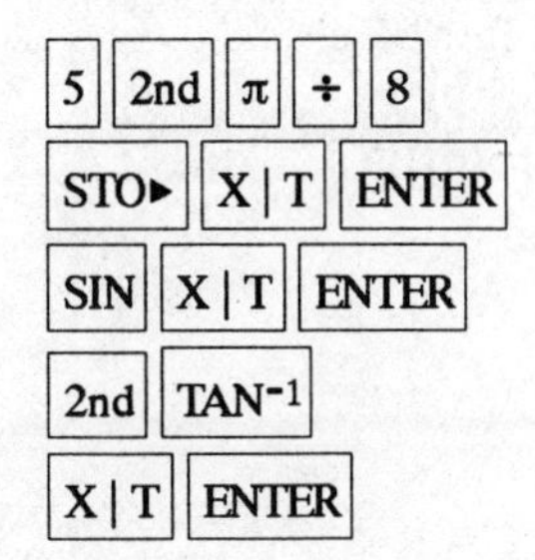

5 2nd π ÷ 8

STO► X | T ENTER

```
5π/8→X
          1.963495408
```

Store $\frac{5\pi}{8}$ as x.

SIN X | T ENTER

```
sin X
          .9238795325
```

Enter $f(x)$ and evaluate.

2nd TAN⁻¹ X | T ENTER

```
tan-1 X
          1.099739749
```

Enter $g(x)$ and evaluate.

Example 2 Evaluate $f(x) = \csc x$ at $x = 32° 5' 45''$.

Solution:

Keystrokes

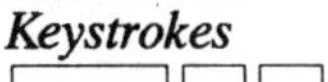

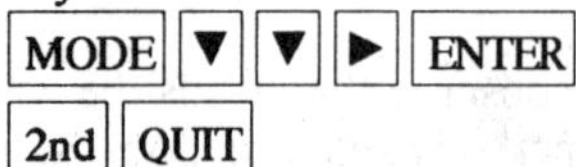

Screen Display

```
Norm Sci Eng
Float 0123456789
Rad Deg
Function Param
Connected Dot
Sequence Simul
Grid Off Grid On
Rect Polar
```

Explanation

The angle measure is given in degrees. Set the calculator for degree measure before starting calculations. Return to the Home screen using [2nd] [QUIT].

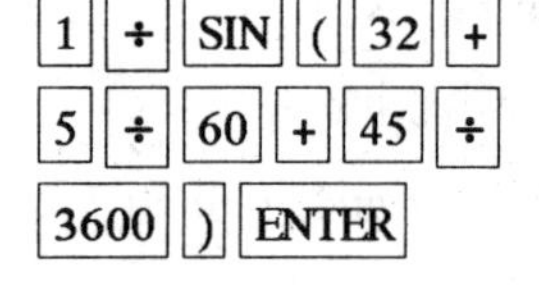

```
1/sin (32+5/60+4
5/3600)
     1.882044822
```

Use $1/\sin x$ as $\csc x$.

Change the minutes and seconds to decimal while entering the angle measure.

Example 3 Graph f(x) = 1.5 sin 2x.

Solution:

Keystrokes

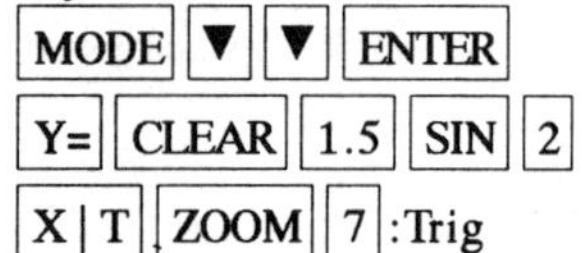

Screen Display

:Y1=1.5sin 2x

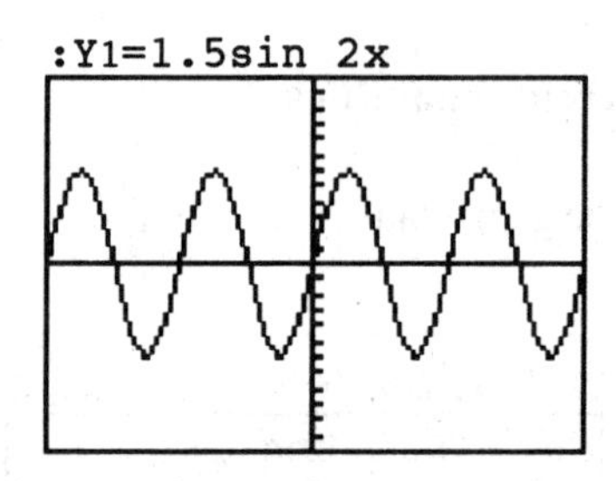

Explanation

Set MODE to Radian measure. Store $f(x)$. Use the trigonometric option on the ZOOM menu to get tick marks set at radian measures on the horizontal axis since the angle measure is in radians. Press [RANGE] to see the RANGE is [-6.28, 6.28]1.57 by [-3,3].25 on the calculator.

Example 4 Graph $g(x) = 3\tan^{-1}(.2x)$.

Solution:

Keystrokes

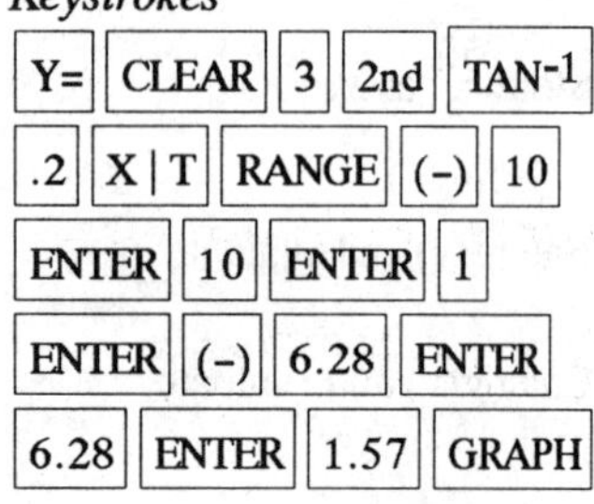

Screen Display

$:Y1=3\tan^{-1}.2X$

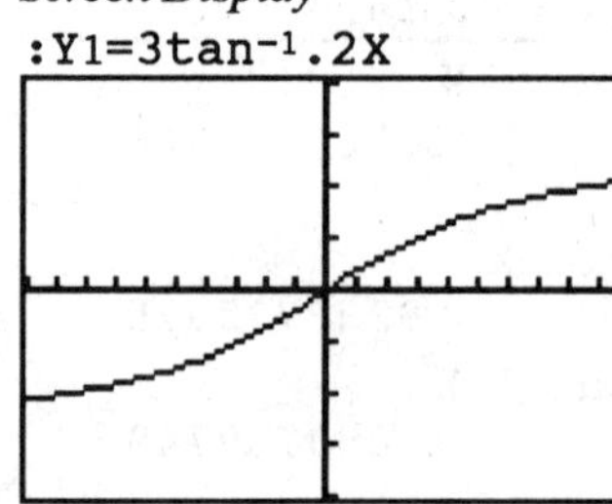

Explanation

Set the RANGE at [-10,10]1 by [-6.28,6.28]1.57

A-19 Polar Coordinates

Example 1 Change the rectangular coordinates $(-\sqrt{3},5)$ to polar form with $r \geq 0$ and $0 \leq \theta \leq 2\pi$.

Solution:

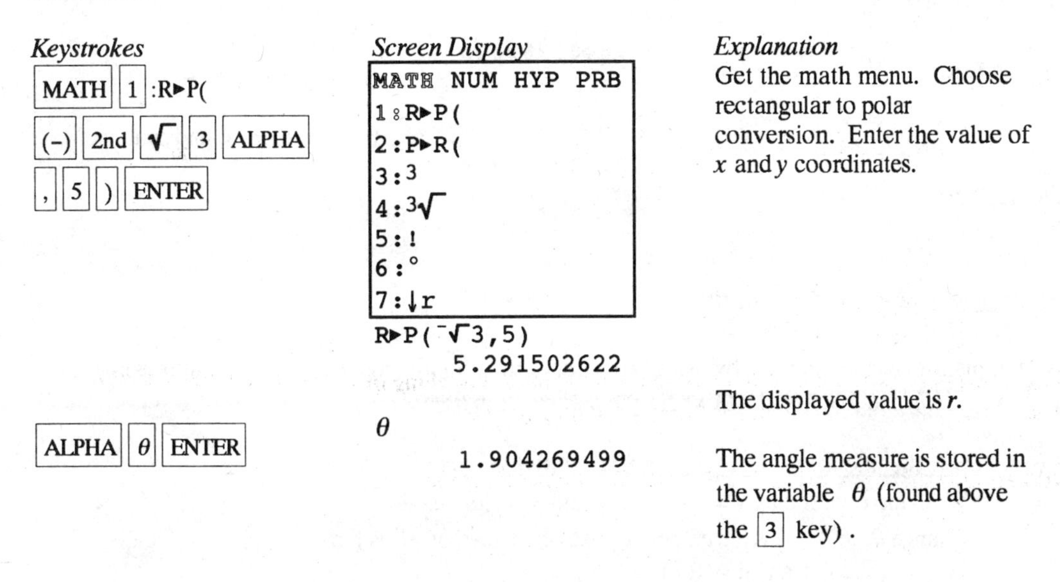

Explanation
Get the math menu. Choose rectangular to polar conversion. Enter the value of x and y coordinates.

The displayed value is r.

The angle measure is stored in the variable θ (found above the 3 key).

Example 2 Change the polar coordinates $(5,\pi/7)$ to rectangular coordinates.

Solution:

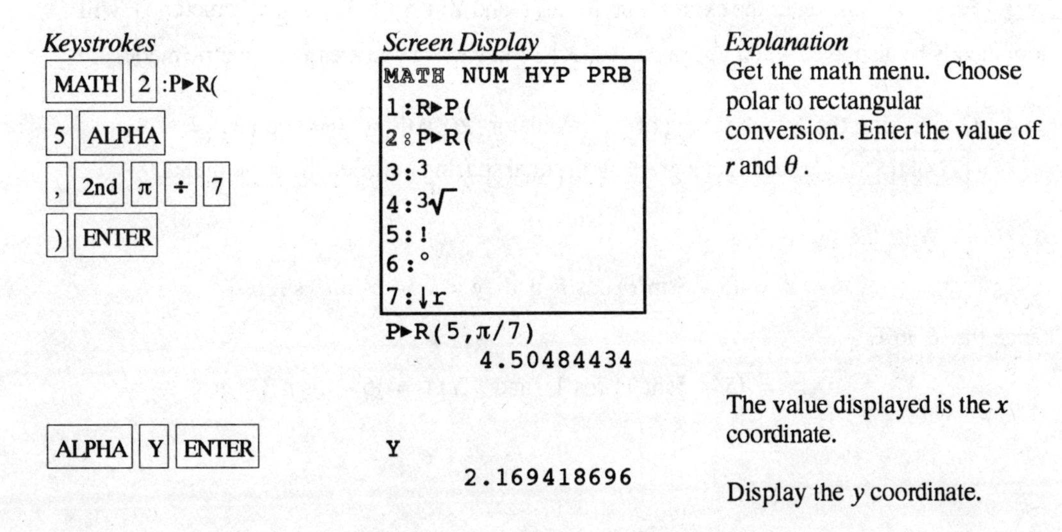

Explanation
Get the math menu. Choose polar to rectangular conversion. Enter the value of r and θ.

The value displayed is the x coordinate.

Display the y coordinate.

Example 3 Evaluate $r = 5 - 5\sin\theta$ at $\theta = \frac{\pi}{7}$.

Solution:

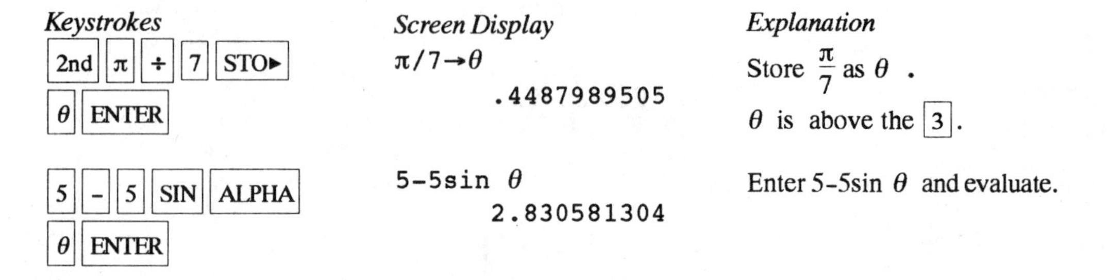

Keystrokes	*Screen Display*	*Explanation*
2nd π ÷ 7 STO▸ θ ENTER	π/7→θ .4487989505	Store $\frac{\pi}{7}$ as θ. θ is above the 3.
5 - 5 SIN ALPHA θ ENTER	5-5sin θ 2.830581304	Enter $5\text{-}5\sin\theta$ and evaluate.

Example 4 Graph $r = 5 - 5\sin\theta$

Polar equations can be graphed by using the parametric graphing mode of the calculator using $x = f(\theta)\cos(\theta)$ and $y = f(\theta)\sin(\theta)$ where the parameter is θ.

In general the steps to graph a polar function are:

Step 1 Write the function as $x = f(\theta)\cos(\theta)$ and $y = f(\theta)\sin(\theta)$.
Change θ to T since the calculator uses T instead of θ as the parametric variable.
So $x = f(T)\cos T$ and $y = f(T)\sin T$.

Step 2 Press MODE, and select parametric mode. Press ENTER.
(See Section A-2 of this manual.)

Step 3 Press Y= and enter the expressions for X1T and Y1T with T as the parameter. T will automatically be displayed when you press the X|T key when you are in parametric mode.

Step 4 Graph using the default setting of the calculator ZOOM 6 :Standard and ZOOM 5 :Square to get a graph with equal spacing between the scale marks.

Solution: Write the function as

$$x = (5 - 5\sin\theta)\cos\theta \text{ and } y = (5 - 5\sin\theta)\sin\theta$$

Change the θ to T

$$X1T = (5 - 5\sin T)\cos T \text{ and } Y1T = (5 - 5\sin T)\sin T$$

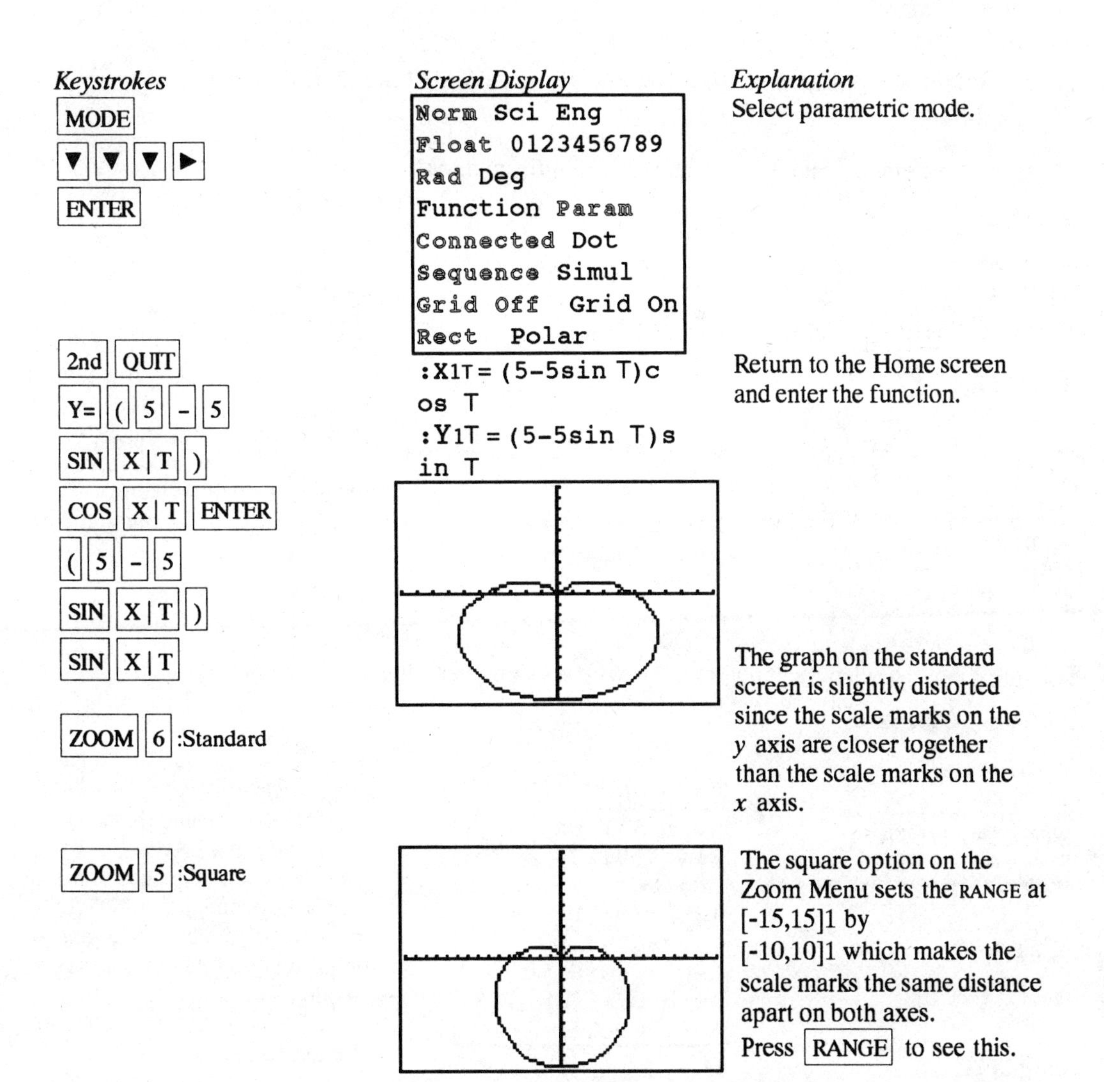

Keystrokes	*Screen Display*	*Explanation*
MODE ▼ ▼ ▼ ► ENTER	Norm Sci Eng Float 0123456789 Rad Deg Function Param Connected Dot Sequence Simul Grid Off Grid On Rect Polar	Select parametric mode.
2nd QUIT Y= (5 - 5 SIN X|T) COS X|T ENTER (5 - 5 SIN X|T) SIN X|T	:X1T= (5-5sin T)c os T :Y1T= (5-5sin T)s in T	Return to the Home screen and enter the function.
ZOOM 6 :Standard		The graph on the standard screen is slightly distorted since the scale marks on the *y* axis are closer together than the scale marks on the *x* axis.
ZOOM 5 :Square		The square option on the Zoom Menu sets the RANGE at [-15,15]1 by [-10,10]1 which makes the scale marks the same distance apart on both axes. Press RANGE to see this.

A-20 Scientific Notation, Significant Digits, and Fixed Number of Decimal Places

Numbers can be entered into the calculator in scientific notation.

Example 1 Calculate $(-8.513\times10^{-3})(1.58235\times10^{2})$. Enter numbers in scientific notation.

Solution:

Keystrokes	*Screen Display*	*Explanation*
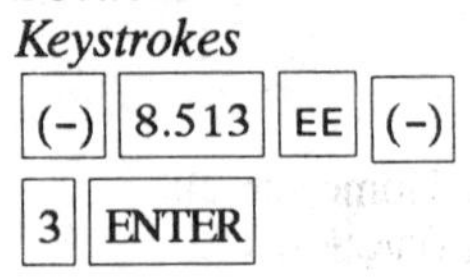	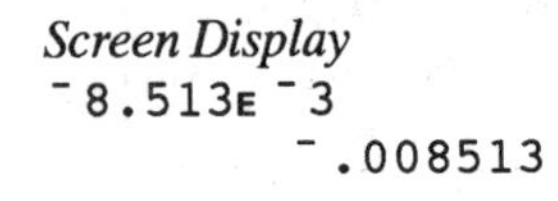	Enter the first number. The number displayed is not in scientific notation. (It is not necessary to press ENTER at this point. This is illustrated to show how the numbers are displayed on the screen.)
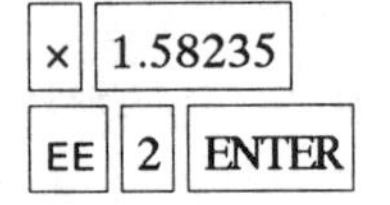	Ans*1.58235E 2 ⁻1.347054555	Multiply by the second number.

Example 2 Set the scientific notation mode with six significant digits and calculate $(351.892)(5.32815\times10^{-8})$.

Solution:

Keystrokes	*Screen Display*	*Explanation*
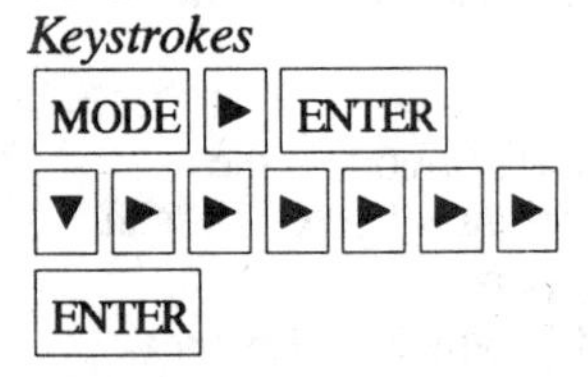	Norm Sci Eng Float 0123456789 Rad Deg Function Param Connected Dot Sequence Simul Grid Off Grid On Rect Polar	Select `Sci` using the arrow keys and press ENTER. Select 5 decimal places using the arrow keys and press ENTER. Five decimal places will give six significant digits in scientific mode.
2nd QUIT		Return to the Home screen.
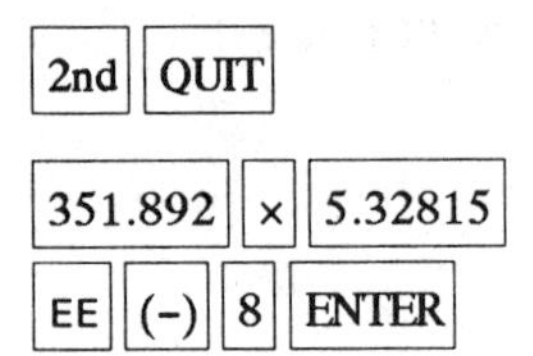	351.892*5.32815E ⁻8 1.87E⁻5	Enter the numbers. Note the result is displayed in scientific notation with six significant digits.

Example 3 Fix the number of decimal places at 2 and calculate the interest earned on $53,218.00 in two years when invested at 5.21% simple interest.

Solution:

Keystrokes	*Screen Display*	*Explanation*
MODE ENTER ▼ ► ► ► ENTER	Norm Sci Eng Float 0123456789 Rad Deg Function Param Connected Dot Sequence Simul Grid Off Grid On Rect Polar	Choose normal notation with 2 fixed decimal points.
2nd QUIT		Return to the Home Screen.
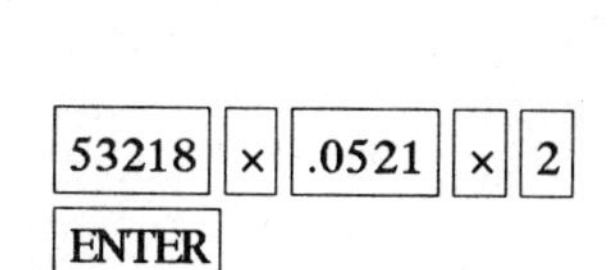	53218*.0521*2 5545.32	Only two decimal places are shown in the answer. The interest is $5545.32.

NOTES

APPENDIX B

CASIO *fx-7700G* GRAPHING CALCULATOR BASIC OPERATIONS

B-1 Getting Started

Press [AC] to turn on the calculator.
A screen similar to the one shown at the right will be displayed. It may be different. Do not be concerned. It is all a matter of how the modes are set.
[SHIFT] [AC] will turn the calculator off.

```
   RUN / COMP
G-type : REC/CON
angle  : Rad
display:Nrml
```

B-2 Calculator Operation

[MODE]

Press [MODE]. There are four boxes displayed on the screen. Each is described below.

```
Sys mode    Cal mode
1:RUN       +:COMP
2:WRT       -:BASE-N
3:PCL       ×:SD
REG model   ÷:REG
4:LIN       Ø:MATRIX
5:LOG       Contrast
6:EXP       ←:LIGHT
7:PWR       →:DARK
```

<u>Sys mode</u>

1:RUN	For normal calculation operations
2:WRT	For writing & checking programs
3:PCL	For clearing existing programs

<u>REG model</u>	(For statistical calculations)
4:LIN	Linear regression model
5:LOG	Logarithmic regression model
6:EXP	Exponential regression model
7:PWR	Power regression model

<u>Cal mode</u>

+:COMP	For normal calculations
-:BASE-N	Binary, octal & hexadecimal calculations
×:SD	For standard deviation calculations
÷:REG	For regression calculations
Ø:MATRIX	For matrix calculations

<u>Contrast</u>

←:LIGHT	To lighten the contrast
→:DARK	To darken the contrast

There is a second menu of modes. Press [AC] [MODE] [SHIFT]. You will have another display of boxes. Each of these is described below.

<u>Stat Data</u>

1:STO	To store statistical data
2:NON-	Not storing statistical data

<u>Stat graph</u>

3:DRAW	To draw a statistical graph
4:NON-	Not drawing statistical graphs

```
Stat Data  Graph type
1:STO      +:REC
2:NON-     -:POL
Stat graph ×:PARAM
           ÷:INEQ
3:DRAW     Draw type
4:NON-     5:CONNECT
           6:PLOT
```

<u>Graph type</u>

+:REC	Rectangular coordinates
−:POL	Polar coordinates
×:PARAM	Parametric graphs
÷:INEQ	Graphing inequalities

<u>Draw type</u>

5:CONNECT	Points connected on graphs
6:PLOT	Plot individual points (unconnected graphs)

Press [MODE] [1] and then [MODE] [+].

Press [MODE] [SHIFT] [+] and [MODE] [SHIFT] [5]. Now you should have the display shown in the box at the top of the previous page, except for the angle measure. The angle could be either Rad or Deg.

This screen says that the calculator is set in:

RUN mode (not a programming mode)
COMP mode (for computations)
REC mode for graphing in rectangular coordinates
CON mode for connected graphs
Rad mode for radian angle measure, Deg for degree measure
Nrm1 mode means that numbers less than .01 and greater
than 10,000,000,000 will be displayed in scientific notation.

These keystrokes will be included in the examples for Example 1 of Section B-4 and Example 1 of Section B-5 Methods 1 and 2 of this manual. It will be assumed that the calculator is set to these settings in following examples. The example will explain in detail if mode settings other than these are needed for a particular problem.

The mode settings can be checked at any time by pressing [[M] Disp].

[SHIFT]

This key must be pressed to access the operation above and to the left of a key. These operations are a light orange color on the face of the calculator. An inverse video [S] is displayed as the cursor on the screen after [SHIFT] key is pressed. In this manual these functions will be referred to in square boxes, just as if the function was printed on the key cap. For example, [(-)] is the function above the [Ans] key.

ALPHA and A-LOCK

This key must be pressed to access the operation above and to the right of a key. These operations are in a red color on the calculator face. An inverse video A is displayed as the cursor on the screen after the ALPHA key is pressed.

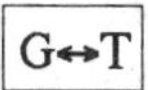

This key switches between the graph screen and the text screen. If you have a graph screen displayed and this key is pressed the text screen will be displayed. If you have a text screen displayed and this key is pressed then the graph screen will be displayed.

Menus

The calculator uses menus for selection of specific functions. The items on the menus are displayed in inverse video across the bottom of the screen. A function key (just below the screen) corresponds to each menu item. Just press the corresponding function key to choose the menu item.

In this manual the menu items will be referred to using the key to be pressed followed by the meaning of the menu. For example, if the RANGE key is pressed the menu item INIT is displayed at the bottom of the screen. Pressing F1 will initialize the RANGE items to a prestored value. In this manual F1 :INIT will refer to this menu item.

PRE

Menus are always displayed across the bottom of the screen. Many of the menu items also have a menu. Hence there are several "layers" of menus. The PRE key is used to get a preceding menu.

Home Screen

The blank screen is called the Home Screen. You can usually get to this screen by pressing AC.

B-3 Correcting Errors

It is easy to correct errors when entering data into the calculator by using the arrow keys, INS, and DEL keys.

◀ or ▶ Moves the cursor to the left or right one position. These keys also replay the most recent entry. ◀ returns the cursor to the rightmost position of the entry. ▶ returns the cursor to the leftmost position of the entry.

▲ Moves the cursor up one line.

▼ Moves the cursor down one line.

DEL Deletes one character at at a time at the cursor position.

SHIFT INS Inserts one or more characters at the cursor position.

B-4 Calculation

Example 1 Calculate $-8 + 9^2 - \left| \frac{3}{\sqrt{2}} - 5 \right|$.

Numbers and characters are entered in the same order as you would read an expression. Do not press [EXE] unless specifically instructed to do so in these examples. Keystrokes are written in a column but you should enter all the keystrokes without pressing the EXE key until [EXE] is displayed in the example.

Solution:

Keystrokes	*Screen Display*	*Explanation*
[MODE] [1]		It is a good idea to check the mode before starting a calculation and to clear the screen. If in doubt set the mode before each calculation. Check mode by pressing [M] Disp.
[MODE] [+]		
[MODE] [SHIFT] [+]		
[MODE] [SHIFT] [5] [AC]		
[SHIFT] [(-)] [8] [+] [9] [x^y] [2] [-]	-8+9x^{y}2-Abs (3÷√	
[SHIFT] [MATH] [F3] :NUM	2-5)	
[F1] :Abs [(] [3] [÷] [√] [2] [-]	70.12132034	
[5] [)] [EXE]		

B-5 Evaluation of an Algebraic Expression

Example 1 Evaluate $\frac{x^4 - 3a}{8w}$ for $x = \pi$, $a = \sqrt{3}$, and $w = 4!$

Two different methods can be used:

1. Store the values of the variables and then enter the expression. When [EXE] is pressed the expression is evaluated for the stored values of the variables.
2. Store the expression and store the values of the variables.
   Recall the expression. Press [EXE]. The expression is evaluated for the stored values of the variables.

The advantage of the second method is that the expression can be easily evaluated for several different values of the variables.

Keystrokes	*Screen Display*
Method 1	
MODE 1 MODE +	
MODE SHIFT +	
MODE SHIFT 5 AC	
SHIFT π → X,θ,T EXE	π→ X 3.141592654
√ 3 → ALPHA A EXE	√3→A 1.732050808
4 SHIFT MATH F2 :PRB	
F1 :x! → ALPHA W EXE	4!→ W 24.

The notation F1 :x! refers to the menu item accessed by pressing the function key F1.

(X,θ,T	
xʸ 4 - 3 ALPHA A) ÷	(Xxʸ4-3A)÷(8W)
(8 ALPHA W) EXE	0.4802757219
Method 2	
MODE 1 MODE +	
MODE SHIFT +	
MODE SHIFT 5 AC	
(X,θ,T xʸ 4 - 3 ALPHA A)	
÷ (8 ALPHA W) [Note: Use ◀ or ▶ to see the entire function.]	
SHIFT F MEM F1 :STO 1 AC	f1:(Xxʸ4-3A)÷(8W
SHIFT π → X,θ,T EXE	π→X 3.141592654
√ 3 → ALPHA A EXE	√3→A 1.732050808
4 SHIFT MATH F2 :PRB F1 :x! → ALPHA W EXE	4!→W 24
SHIFT F MEM F2 :RCL 1	(Xxʸ4-3A)÷(8W)
EXE	0.4802757219

Example 2 For $f(x) = 3x+5$ and $g(x) = \sqrt{x-\sqrt{x}}$ find $f(2) - g(2)$.

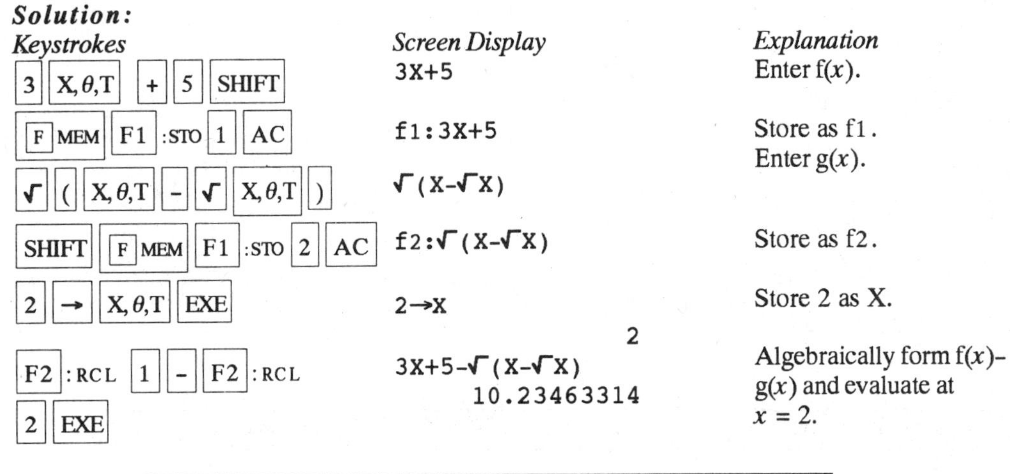

Solution:

Keystrokes	*Screen Display*	*Explanation*
3 X,θ,T + 5 SHIFT	3X+5	Enter f(x).
F MEM F1 :STO 1 AC	f1:3X+5	Store as f1. Enter g(x).
√ (X,θ,T - √ X,θ,T)	√(X-√X)	
SHIFT F MEM F1 :STO 2 AC	f2:√(X-√X)	Store as f2.
2 → X,θ,T EXE	2→X 2	Store 2 as X.
F2 :RCL 1 - F2 :RCL 2 EXE	3X+5-√(X-√X) 10.23463314	Algebraically form f(x)-g(x) and evaluate at x = 2.

Check the mode settings by pressing M Disp .
(See Section B-2 above on how to set the modes.)

B-6 Testing Inequalities in One Variable

Example 1 Determine whether or not $x^3 + 5 < 3x^4 - x$ is true for $x = -\sqrt{2}$.

Solution:

Keystrokes	*Screen Display*	*Explanation*
SHIFT (-) √ 2 → X,θ,T EXE	⁻√2→X -1.414213562	Store the value for x.
X,θ,T xʸ 3 + 5 SHIFT F MEM F1 :STO 1	f1:Xxʸ3+5	Enter and store the left side of the expression.
AC 3 X,θ,T xʸ 4 - X,θ,T F1 :STO 2	f2:3Xxʸ4-X	Enter and store the right side of the expression.
F2 :RCL 1 EXE	Xxʸ3+5 2.171572875	Recall the first expression and evaluate.
F2 :RCL 2 EXE	3Xxʸ4-X 13.41421356	Recall the expression and evaluate.

Comparing of the values of these functions shows that the inequality expression is true for this value of x.

B-7 Graphing and Setting the Range

Up to six functions can be stored in the calculator one time. Functions are graphed one at a time with the previous functions remaining on the screen until the screen is cleared.

Example 1 Graph $y = x^2$, $y = .5x^2$, $y = 2x^2$, and $y = -1.5x^2$ on the same coordinate axes.

Solution:

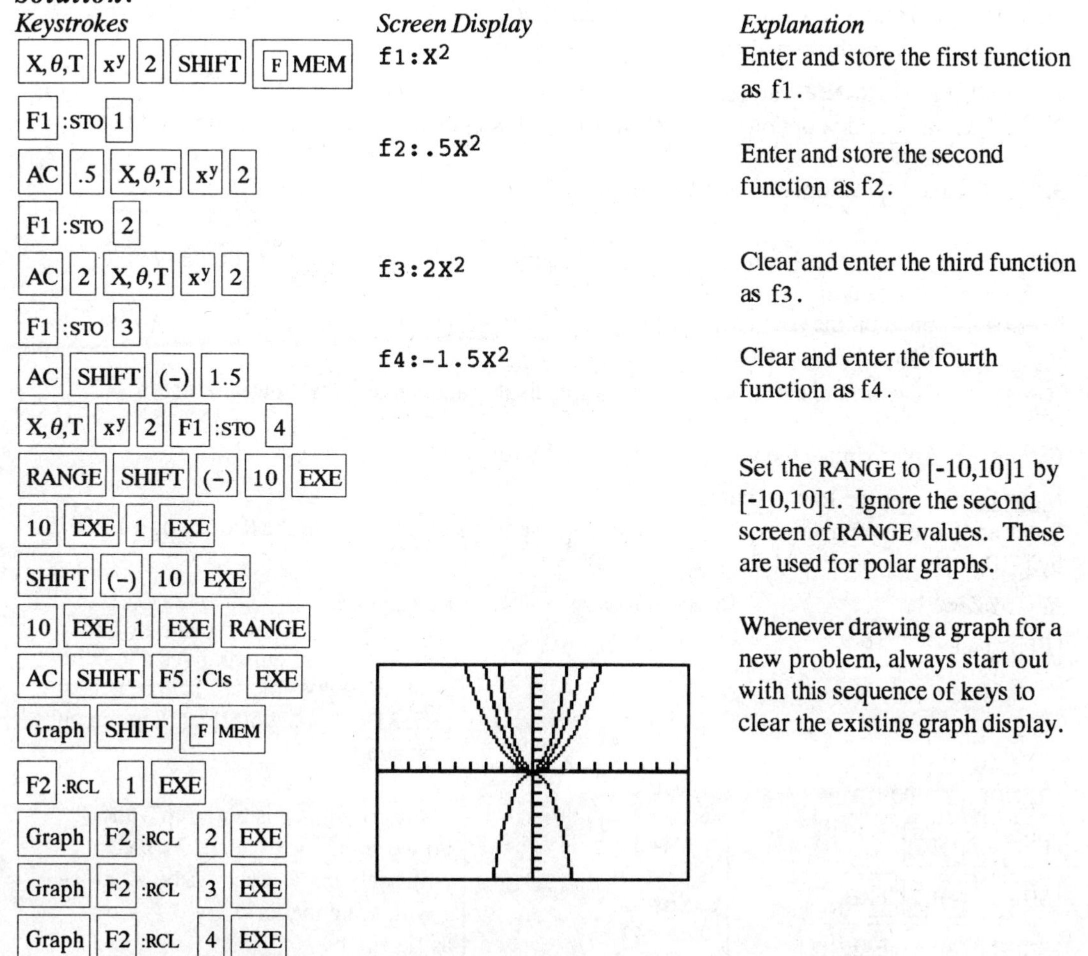

Keystrokes	*Screen Display*	*Explanation*
X,θ,T xʸ 2 SHIFT F MEM	f1:X2	Enter and store the first function as f1.
F1 :STO 1		
AC .5 X,θ,T xʸ 2	f2:.5X2	Enter and store the second function as f2.
F1 :STO 2		
AC 2 X,θ,T xʸ 2	f3:2X2	Clear and enter the third function as f3.
F1 :STO 3		
AC SHIFT (-) 1.5	f4:-1.5X2	Clear and enter the fourth function as f4.
X,θ,T xʸ 2 F1 :STO 4		
RANGE SHIFT (-) 10 EXE		Set the RANGE to [-10,10]1 by [-10,10]1. Ignore the second screen of RANGE values. These are used for polar graphs.
10 EXE 1 EXE		
SHIFT (-) 10 EXE		
10 EXE 1 EXE RANGE		Whenever drawing a graph for a new problem, always start out with this sequence of keys to clear the existing graph display.
AC SHIFT F5 :Cls EXE		
Graph SHIFT F MEM		
F2 :RCL 1 EXE		
Graph F2 :RCL 2 EXE		
Graph F2 :RCL 3 EXE		
Graph F2 :RCL 4 EXE		

[-10,10]1 by [-10,10]1 means to set the RANGE for the graph at $-10<x<10$ and $-10<y<10$ with scale marks on the axes at every unit.

SHIFT F5 :CLS EXE will clear the graph screen.

B-8 TRACE, ZOOM, RANGE and G↔T

[F1] :TRACE allows you to observe both the x and y coordinate of a point on the graph when you use [◄] or [►] to move the cursor along the graph. The cursor will move along the most recently graphed function. The calculator takes about 10 seconds to set up after pressing [F1].

[F2] :ZOOM will magnify a graph so the coordinates of a point can be approximated with greater accuracy. There are three methods to zoom in:

1. Change the RANGE values.
2. Use the [F1] :BOX option on the zoom menu in conjunction with [F2] :FCT option to set the factors.
3. Use the [F3] :xf option on the zoom menu.

[F4] :x'/f option on the zoom menu is used to zoom out. (See Example 1 Section B-9 of this manual).

[F5] :ORG option on the zoom menu will redraw the most recent function graphed using your original RANGE settings.

[G↔T] allows you to move between the graph display and the text (calculation) screen.

Example 1 Approximate the value of x if $y = -1.58$ for $y = x^3 - 2x^2 + \sqrt{x} - 8$.

Method 1 Change the RANGE values.
Enter and store the function as f1. Clear the graph screen and graph using the RANGE of [-10,10]1 by [-10,10]1 (See Section B-7 of this manual).

Keystrokes	*Screen Display*	*Explanation*
[F1] :TRACE [►] ... [►]		Press the right arrow repeatedly until the new type of cursor gives a y value as close as possible to -1.58 which is (2.5531914, -2.795991). (Diagram on next age.)

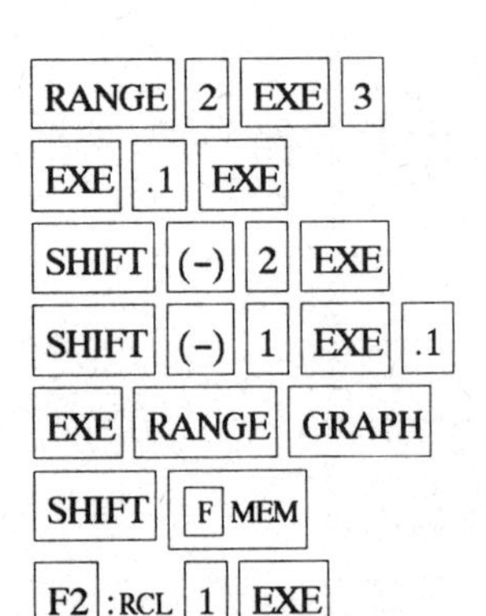


```
RANGE
Xmin=2
 max=3
 scl=.1
Ymin=-2
 max=-1
 scl=.1
```

The x coordinate is between 2 and 3. So we set the RANGE at $2<x<3$ with scale marks every .1 by $-2<y<-1$ with scale marks every .1 (written as [2,3].1 by [-2,-1].1.

[F1] :TRACE can be used again to estimate a new x value. Repeat using trace and changing the RANGE until the approximation of (2.67,-1.58) has been found.

Method 2 Use the F4 :xf option on the F2 :Zoom menu.

Enter and store the function as f1. Clear the graph screen and graph using the RANGE of [-10,10]1 by [-10,10]1 (See Section B-7 of this manual).

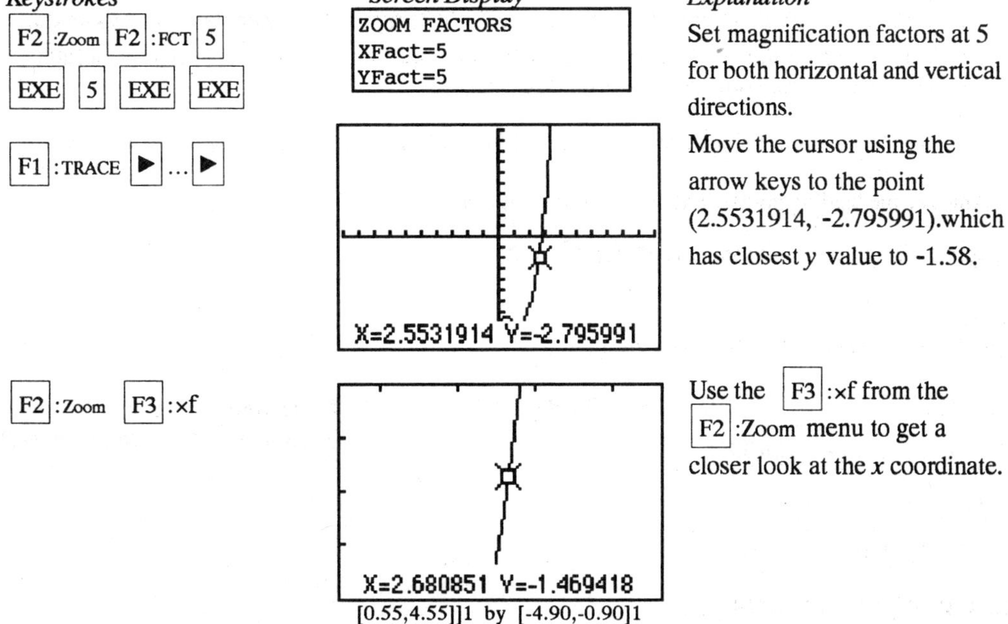

Explanation

Set magnification factors at 5 for both horizontal and vertical directions.

Move the cursor using the arrow keys to the point (2.5531914, -2.795991).which has closest y value to -1.58.

Use the F3 :xf from the F2 :Zoom menu to get a closer look at the x coordinate.

Repeat this procedure until you get a value for the x coordinate accurate to two decimal places. The point has coordinates (2.67,-1.58).

Method 3 Use the F1 :BOX option on the F2 :Zoom menu.

Enter and store the function as f1. Graph using the RANGE of [-10,10]1 by [-10,10]1 (See Section B-7 of this manual).

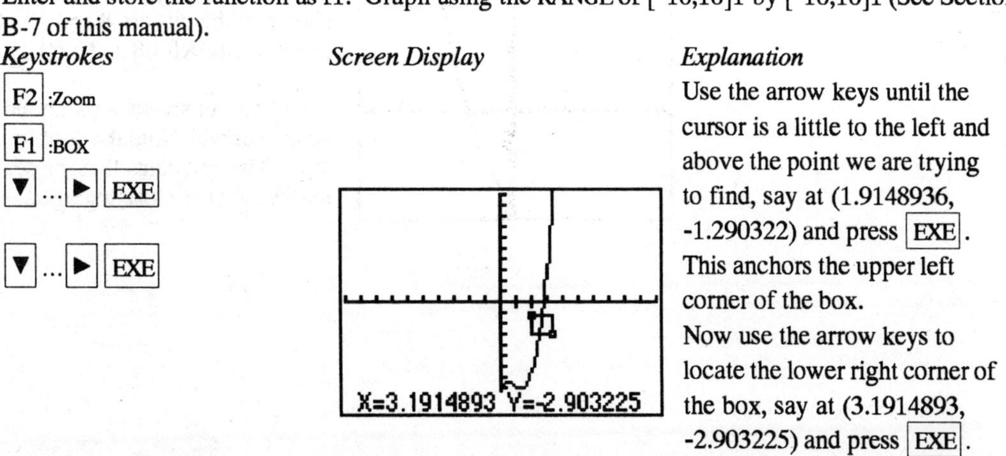

Explanation

Use the arrow keys until the cursor is a little to the left and above the point we are trying to find, say at (1.9148936, -1.290322) and press EXE. This anchors the upper left corner of the box.

Now use the arrow keys to locate the lower right corner of the box, say at (3.1914893, -2.903225) and press EXE.

Repeat this procedure until you get a value for the x coordinate accurate to two decimal places. The point has coordinates (2.67,-1.58).

B-9 Determining the Range

There are several ways to determine the RANGE of the values that should be used for the limits of the x and y axes. Three are described below:

1. Graph using the default setting F1 :INIT on the RANGE menu of the calculator and zoom out. The disadvantage of this method is that often the function cannot be seen at either the default settings or the zoomed out settings of the RANGE.
2. Evaluate the function for several values of x. Make a first estimate based on these values.
3. Analyze the leading coefficient and the constant terms.

Example 1 Graph the function $f(x) = .2x^2 + \sqrt[3]{x} - 32$.

Solution:

Method 1 Use the default setting and zoom out.

Keystrokes | *Screen Display* | *Explanation*

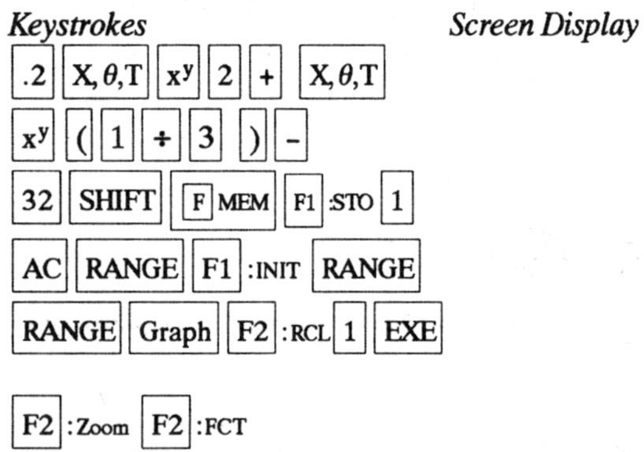

Enter and store the function as f1.

Graph using the initialized screen. Nothing is seen on the graph screen because no part of this curve is in the initialized RANGE.

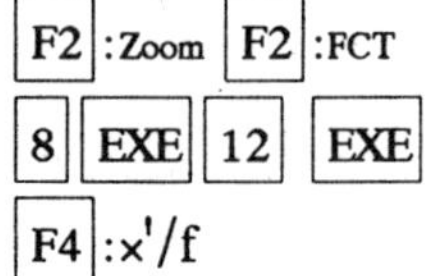

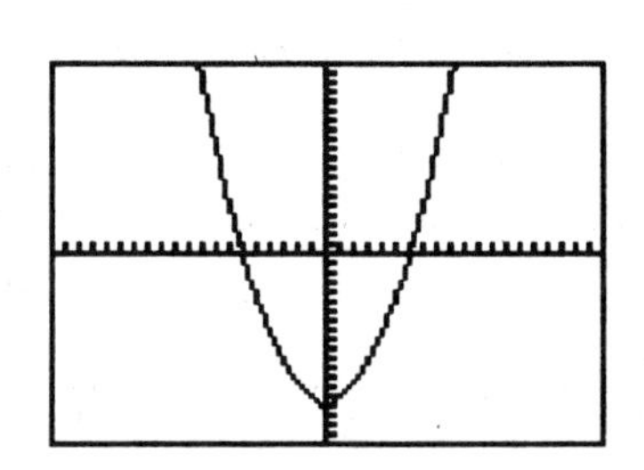

Factors used to get the graph shown at the left are 8 and 12. In other words, Xfct:8 and Yfct:12.

Zooming out shows a parabolic shaped curve. Note the double axis. This indicates that the scale marks are very close together.

Method 2 Evaluate the function for several values of x. One decimal place accuracy is enough for approximating the RANGE. (See Section B-5 of this manual on how to evaluate a function at given values of x.)

x	f(x)
-20	45.3
-10	-14.2
0	-32.0
10	-9.8
20	-50.7

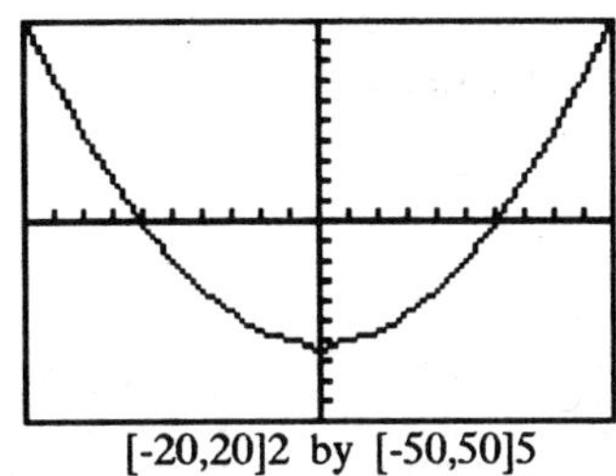
[-20,20]2 by [-50,50]5

Analyzing this table indicates that a good RANGE to start with is [-20,20]2 by [-50,50]5.

Method 3 Analyze the leading coefficient and constant terms.
Since the leading coefficient is .2 the first term will increase 2 units for each 10 units x^2 increases ($\sqrt{10}$ or about 3 units increase in x). A first choice for the x-axis limits can be found using:

$\frac{10\times(\text{unit increase in } x)}{(\text{first term increase})} = \frac{10\times3}{2} = 15$. Start by setting Xmin=-15 and Xmax=15.

A first choice for the scale having about 20 marks on the axis can be found using $\frac{\text{Xmax-Xmin}}{20} =$

$\frac{15-(-15)}{20} = 1.5$ (round to 2). So the limits on the x axis could be [-15,15]2.

A first choice for the y-axis limits could be ±(constant term). The scale can be found using

$\frac{\text{Ymax-Ymin}}{20} = \frac{32-(-32)}{20} = 3.2$ (round to 4).

So a first choice for the y-axis limits could be [-32,32]4.

Set the RANGE to [-15,15]2 by [-32,32]4.

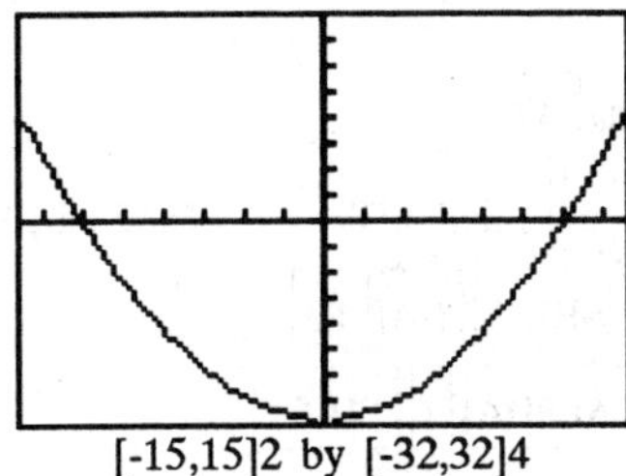
[-15,15]2 by [-32,32]4

B-10 Piecewise-Defined Functions

There are two methods to graph piecewise-defined functions:

1. Graph each piece of the function separately as an entire function on the same coordinate axes. Use trace and zoom to locate the partition value on each of the graphs.
2. Store each piece of the function separately but include an inequality statement following the expression which will set the RANGE of values on x for which the function should be graphed. Then graph all pieces on the same coordinate axes.

Example 1 Graph $f(x) = \begin{cases} x^2+1 & x < 1 \\ 3x-5 & x \geq 1 \end{cases}$

Method 1

Keystrokes	*Screen Display*	*Explanation*
SHIFT F5 :Cls AC		Enter the functions.
X,θ,T x^y 2 + 1 SHIFT	Xx^{y}2+1	Graph. Both functions will be displayed. Use trace and zoom to find the point on the graphs where x=1.
F MEM F1 :STO 1 AC	f1:Xx^{y}2+1	
3 X,θ,T - 5 SHIFT	3X-5	
F MEM F1 :STO 2 AC	f2:3X-5	
RANGE SHIFT (-) 10	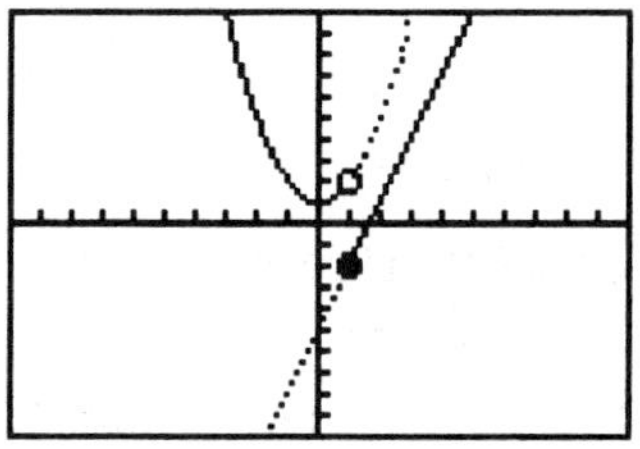	The graph shown to the left shows the curves with the partition values indicated. This result cannot be displayed on the calculator; it must be added by hand.
EXE 10 EXE 1 EXE		
SHIFT (-) 10 EXE 10		
EXE 1 EXE RANGE		
Graph SHIFT F MEM		
F2 :RCL 1 EXE		
AC Graph SHIFT		
F MEM F2 :RCL 2 EXE		

Method 2

Keystrokes	*Screen Display*	*Explanation*
SHIFT F5 :Cls AC		The square bracketed expression will yield a 1 when $-10<x<1$ and a 0 elsewhere. Hence the first function stored will be graphed only for $-10<x<1$.
X,θ,T x^y 2 + 1 SHIFT	Xx^{y}2+1,[-10,1]	
, ALPHA [SHIFT (-) 10		
SHIFT , 1 ALPHA] SHIFT		
F MEM F1 :STO 1 AC	f1:Xx^{y}2+1,[-10,1]	
3 X,θ,T - 5 SHIFT ,	3X-5,[1,10]	
ALPHA [1 SHIFT ,	f2:3X-5,[1,10]	Similarly, the second function stored will be graphed only for $1<x<10$.
10 ALPHA] SHIFT	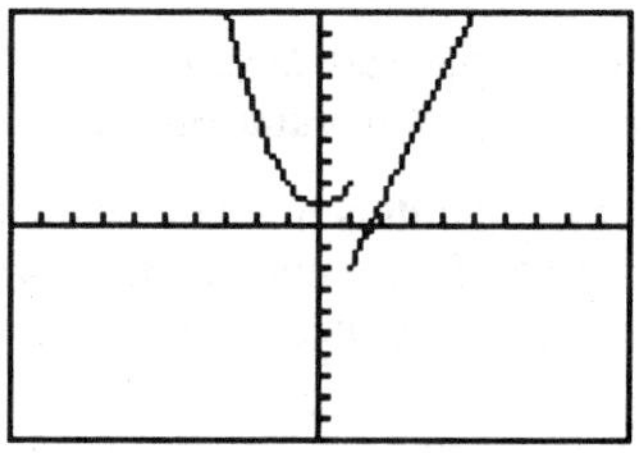	
F MEM F1 :STO 2 AC		
GRAPH SHIFT F MEM		
F2 :RCL 1 EXE		
AC Graph SHIFT		
F MEM F2 :RCL 2 EXE		

B-11 Solving Equations in One Variable

There are two methods for approximating the solution of an equation using graphing.

1. Set the expression equal to zero. Graph y=(the expression). Find where the curve crosses the x axis. The x values (the x intercepts) are the solutions to the equation.

2. Graph y = left side of the equation) and y = (right side of the equation) on the same coordinate axes. The x coordinate of the point(s) of intersection are the solutions to the equation.

<u>Example 1</u> Solve $\dfrac{3x^2}{2} - 5 = \dfrac{2(x+3)}{3}$.

<u>Method 1</u> Write the equation as $\left(\dfrac{3x^2}{2} - 5\right) - \left(\dfrac{2(x+3)}{3}\right) = 0$. Graph $y = \left(\dfrac{3x^2}{2} - 5\right) - \left(\dfrac{2(x+3)}{3}\right)$ and find the x value where the graph crosses the x axis. This is the x intercept.

Keystrokes	*Screen Display*	*Explanation*
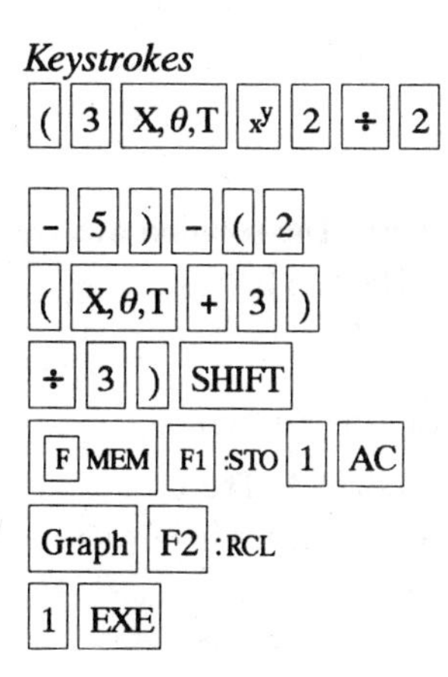	(3Xxʸ2÷2−5)−(2(X +3)÷3)	Enter the expression and store as f1.
	f1:(3Xxʸ2÷2−5)−(	Use trace and zoom to find the x intercepts. They are: $x \approx$-1.95 and $x \approx$2.39.
	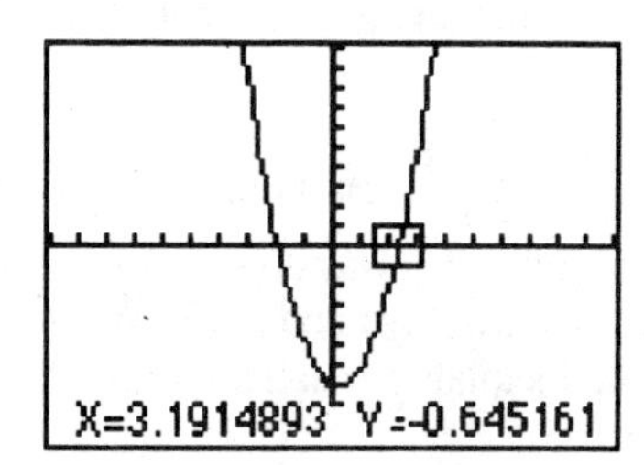	Hence the approximate solutions to this equation are -1.95 and 2.39.

<u>Method 2</u> Graph $y = \dfrac{3x^2}{2} - 5$ and $y = \dfrac{2(x+3)}{3}$ on the same coordinate axes and find the x coordinate of their points of intersection.

Keystrokes

3 X,θ,T xʸ 2 ÷ 2 -
5 SHIFT F MEM F1 :STO
1 AC 2 (X,θ,T + 3
) ÷ 3 SHIFT F MEM
F1 :STO 2 AC Graph SHIFT
F MEM F2 :RCL 1 SHIFT
PRGM F6 :: Graph SHIFT
F MEM F2 :RCL 2 EXE

Screen Display

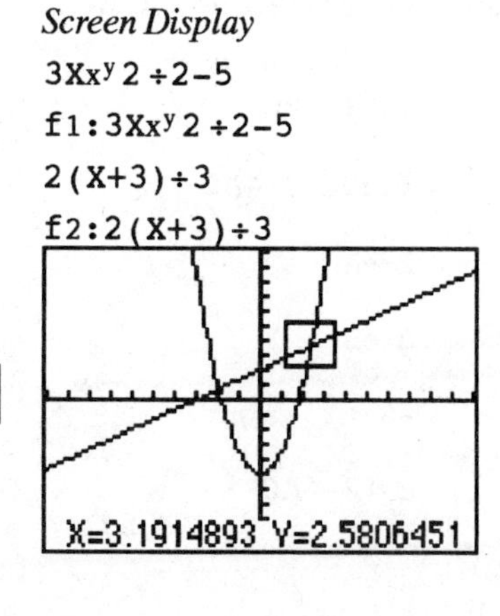

Explanation

Enter the two functions.

Use trace and zoom to find the x values of the points of intersection. The x values are: $x \approx$-1.95 and $x \approx$2.39.

Note: Use a colon between the Graph commands to automatically graph both curves after a zoom.

Hence the approximate solutions to this equation are -1.95 and 2.39.

B-12 Solving Inequalities in One Variable

There are two methods for approximating the solution of an inequality using graphing.

1. Write the inequality with zero on one side of the inequality sign and the expression on the other side. Graph y = (the expression). Find where the curve crosses the x axis. The solution will be an inequality with the x value as the cut off numbers.
2. Graph y = (left side of the inequality) and y=(right side of the inequality) on the same coordinate axes. The x coordinate of the points of intersection are the solutions to the equation. Identify which side of the x value satisfies the inequality by observing the graphs of the two functions.

Example 1 Approximate the solution to $\frac{3x^2}{2} - 5 \le \frac{2(x+3)}{3}$ using two decimal places.

Solution:

Method 1 Write the equation as $\left(\frac{3x^2}{2} - 5\right) - \left(\frac{2(x+3)}{3}\right) \le 0.$

Graph $y = \left(\frac{3x^2}{2} - 5\right) - \left(\frac{2(x+3)}{3}\right)$ and find the x intercept(s).

This was done in Example B-6 Method 1. The x intercepts are -1.95 and 2.39. The solution to the inequality is the interval on x for which the graph is below the x axis. The solution is $-1.95 \le x \le 2.39$.

Method 2 Graph $y = \frac{3x^2}{2} - 5$ and $y = \frac{2(x+3)}{3}$ on the same coordinate axes and find the x coordinate of their points of intersection. This was done in Example B-6 Method 2. The x coordinate of the points of intersections are -1.95 and 2.39. We see that the parabola is below the line for $-1.95 \le x \le 2.39$. Hence the inequality is satisfied for $-1.95 \le x \le 2.39$.

B-13 Storing an Expression That Will Not Graph

None of the functions stored will be graphed unless the command using the key [Graph] is used. Any expression can be stored as a function and later evaluated for stored values of the variables.

Example 1 Store the expression B^2-4AC and evaluate it at A=2, B=3, and C=-6.

Keystrokes	*Screen Display*		*Explanation*
[ALPHA] [B] [x^y] [2] [-] [4]	Bx^y2-4AC		Enter and store the expression as f1.
[ALPHA] [A] [ALPHA] [C] [SHIFT]	f1:Bx^y2-4AC		
[F MEM] [F1] :STO [1] [AC]	2→A		
[2] [→] [ALPHA] [A] [EXE]		2.	Store the values of the variables.
[3] [→] [ALPHA] [B] [EXE]	3→B		
		3.	
[SHIFT] [(-)] [6] [→] [ALPHA] [C]	-6→A		
		-6.	Evaluate the expression. The result is 57.
[EXE] [SHIFT] [F MEM]	Bx^y2-4AC		
[F2] :RCL [1] [EXE]		57.	

B-14 Permutations and Combinations

Example 1 Find (A) $P_{10,3}$ and (B) $C_{12,4}$

Solution (A):

Keystrokes	*Screen Display*	*Explanation*
[10] [SHIFT]	10P3	Enter the first number. Enter the math menu and choose PRB by pressing F2. Then choose nPr by pressing F2 a second time. Enter the second number and press [EXE].
[MATH] [F2] :PRB [F2] :nPr	720.	
[3] [EXE]		

Solution (B):

Keystrokes	*Screen Display*	*Explanation*
[12] [SHIFT]	12C4	Enter the first number. Enter the math menu and choose PRB by pressing F2. Then choose nCr by pressing F3. Enter the second number and press [EXE].
[MATH] [F2] :PRB [F3] :nCr	495.	
[4] [EXE]		

B-15 Matrices

Example 1 Given the matrices

$$A = \begin{bmatrix} 1 & -2 \\ 3 & 0 \\ 5 & -8 \end{bmatrix} \quad B = \begin{bmatrix} 2 & 1 & 5 \\ 3 & 2 & -1 \\ 0 & 8 & -3 \end{bmatrix}$$

Find (A) $-3BA$ (B) B^{-1} (C) A^{T} (D) $\det B$

Solution (A):

Keystrokes	*Screen Display*	*Explanation*
[MODE] [0]	MAT A	Enter the matrix mode.
[F1] :A [F6] :▽	Row:3 Colm:2	Choose matrix A and enter dimensions.
[F1] :DIM [3] [EXE] [2] [EXE]	A 1 2	
[1] [EXE] [SHIFT] [(-)]	1 [1 -2]	Enter the matrix A elements.
[2] [EXE] [3] [EXE]	2 [3 0]	
[0] [EXE] [5] [EXE] [SHIFT]	3 [5 -8]	
[(-)] [8] [EXE] [PRE]		

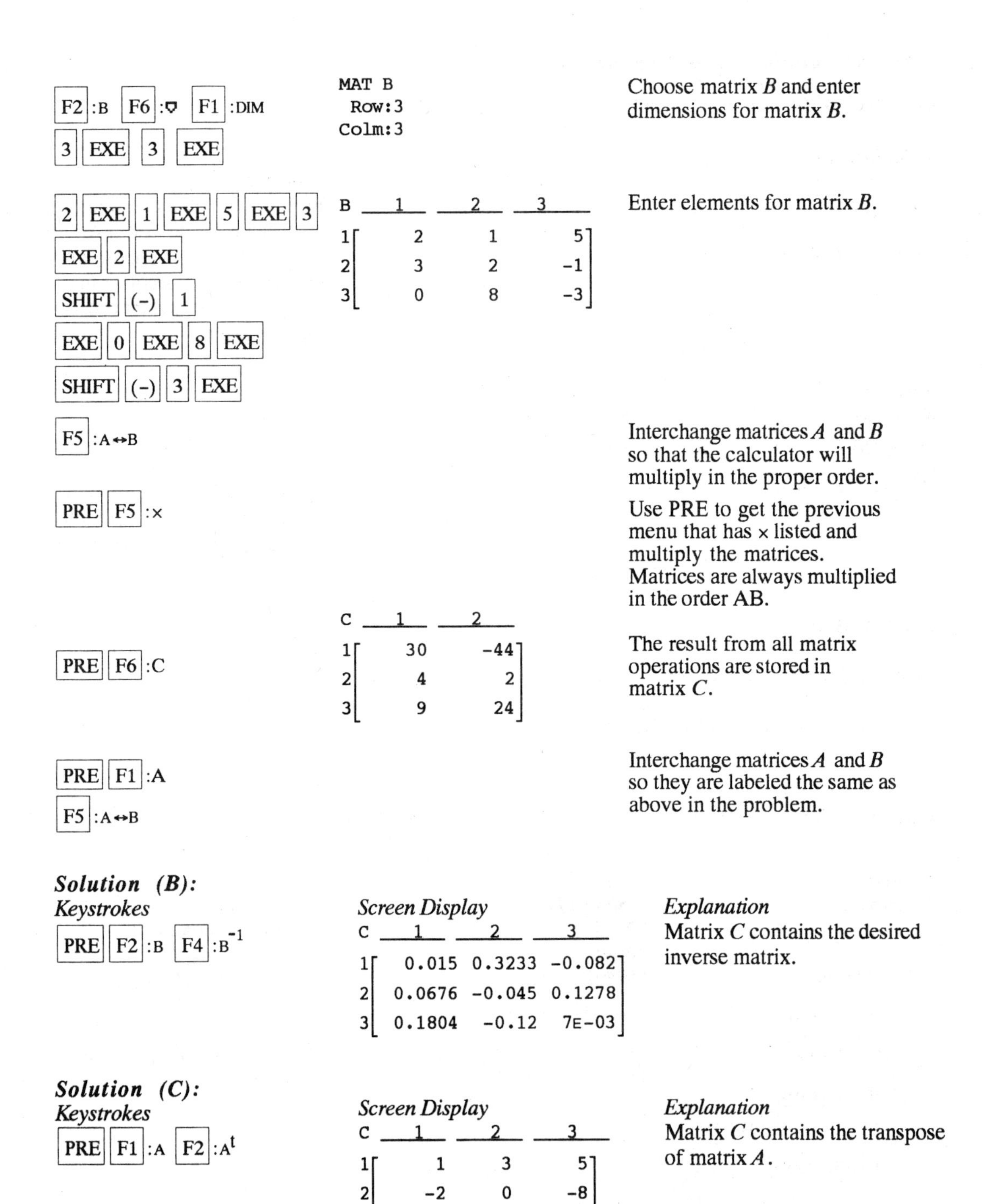

Keystrokes	Screen Display	Explanation
F2 :B F6 :▽ F1 :DIM 3 EXE 3 EXE	MAT B Row:3 Colm:3	Choose matrix B and enter dimensions for matrix B.
2 EXE 1 EXE 5 EXE 3 EXE 2 EXE SHIFT (-) 1 EXE 0 EXE 8 EXE SHIFT (-) 3 EXE	B 1 2 3 1 [2 1 5] 2 [3 2 -1] 3 [0 8 -3]	Enter elements for matrix B.
F5 :A↔B		Interchange matrices A and B so that the calculator will multiply in the proper order.
PRE F5 :×		Use PRE to get the previous menu that has × listed and multiply the matrices. Matrices are always multiplied in the order AB.
PRE F6 :C	C 1 2 1 [30 -44] 2 [4 2] 3 [9 24]	The result from all matrix operations are stored in matrix C.
PRE F1 :A F5 :A↔B		Interchange matrices A and B so they are labeled the same as above in the problem.

Solution (B):

Keystrokes	*Screen Display*	*Explanation*
PRE F2 :B F4 :B⁻¹	C 1 2 3 1 [0.015 0.3233 -0.082] 2 [0.0676 -0.045 0.1278] 3 [0.1804 -0.12 7E-03]	Matrix C contains the desired inverse matrix.

Solution (C):

Keystrokes	*Screen Display*	*Explanation*
PRE F1 :A F2 :Aᵗ	C 1 2 3 1 [1 3 5] 2 [-2 0 -8]	Matrix C contains the transpose of matrix A.

Solution (D):

Keystrokes	*Screen Display*	*Explanation*
PRE F2 :B F3 :|B|	detB = 133.	The determinant of matrix B is 133.

Example 2 Find the reduced form of matrix $\begin{bmatrix} 2 & 1 & 5 & 1 \\ 3 & 2 & -1 & -5 \\ 0 & 8 & -3 & 10 \end{bmatrix}$

Enter the matrix dimensions and elements for matrix A using the procedure in Example 1 of Section B-15. Notice the right bracket is missing which indicates that the entire matrix does not appear on the screen. Use the arrow keys to see the other elements of the matrix.

Solution:

Keystrokes	*Screen Display*	*Explanation*
F1 :A .5 F1 :kA	C: 1 [1 0.5 2.5; 2 [1.5 1 -0.5; 3 [0 4 -1.5	Matrix A must be displayed on the screen. Multiply all elements of matrix A by .5.
F2 :C→B	B: 1 [1 0.5 2.5; 2 [1.5 1 -0.5; 3 [0 4 -1.5	Store the result as matrix B.
		We wish to multiply row 1 of matrix B by -1.5 and add it to row 2. This is done by multiplying on the left by an elementary row matrix which is entered as matrix A. $A=\begin{bmatrix} 1 & 0 & 0 \\ -1.5 & 1 & 0 \\ 0 & 0 & 1 \end{bmatrix}$.
PRE F1 :A F6 :▽ F1 :DIM 3 EXE 3 EXE 1 EXE 0 EXE 0 EXE SHIFT (-) 1.5 EXE 1 EXE 0 EXE 0 EXE 0 EXE 1 EXE PRE	A: 1 [1 0 0]; 2 [-1.5 1 0]; 3 [0 0 1]	Store matrix A.
F5 :×	C: 1 [1 0.5 2.5; 2 [0 0.25 -4.25; 3 [0 4 -1.5	Multiply the matrices to find AB. The result is in matrix C.
F2 :C→B	B: 1 [1 0.5 2.5; 2 [0 0.25 -4.25; 3 [0 4 -1.5	Store the result in matrix B.

PRE F1 :A 1 EXE
SHIFT (-) 2 EXE
0 EXE 0 EXE 4 EXE 0
EXE 0 SHIFT (-) 16
1 EXE

A	1	2	3
1	1	-2	0
2	0	4	0
3	0	-16	1

We now wish to multiply row 2 of matrix B by 4. We could do this by using another elementary row matrix $A = \begin{bmatrix} 1 & 0 & 0 \\ 0 & 4 & 0 \\ 0 & 0 & 1 \end{bmatrix}$. However we can see that we will then have to multiply row 2 by -4 and add to row 3. This means that row 2 will need to be multiplied by 4(-4) now. Also we will want to multiply row 2 by -.5 and add to row 1. All of these operations can be done by using

$$A = \begin{bmatrix} 1 & 4(-.5) & 0 \\ 0 & 4 & 0 \\ 0 & 4(-4) & 1 \end{bmatrix} = \begin{bmatrix} 1 & -2 & 0 \\ 0 & 4 & 0 \\ 0 & -16 & 1 \end{bmatrix}$$

PRE F5 :×

C	1	2	3
1	1	0	11
2	0	1	-17
3	0	0	66.5

Multiply matrices to find AB. The result is in C.

F2 :C→B

B	1	2	3
1	1	0	11
2	0	1	-17
3	0	0	66.5

Store this result in result in matrix B.

Finally we need to get a 1 in row 3 column 3 and use this 1 to get 0's above it. The following elementary row matrix will accomplish this:

$$A = \begin{bmatrix} 1 & 0 & -11(1 \div 66.5) \\ 0 & 1 & 17(1 \div 66.5) \\ 0 & 0 & 1 \div 66.5 \end{bmatrix}$$

$$= \begin{bmatrix} 1 & 0 & -0.1650 \\ 0 & 1 & 0.2556 \\ 0 & 0 & 0.015 \end{bmatrix}$$

PRE F1 :A 1 EXE 0 EXE
SHIFT
(-) 11 (1 ÷ 66.5)
EXE 0 EXE 1 EXE 17 (
1 ÷ 66.5) EXE 0 EXE
0 EXE 1 ÷ 66.5 EXE

A	1	2	3
1	1	0	-0.1650
2	0	1	0
3	0.2556	0	0.015

PRE F5 :×

C	1	2	3
1	1	0	0
2	0	1	-1E-11
3	0	0	.9999

Find AB. The result is in matrix C. Hence the reduced form of the original matrix is:

$$\begin{bmatrix} 1 & 0 & 0 & -2.426 \\ 0 & 1 & 0 & 1.5714 \\ 0 & 0 & 1 & 0.8571 \end{bmatrix}$$

B-16 Graphing an Inequality

There are two ways to graph an inequality.

1. Graph the boundary curve. Determine the half-plane by choosing a test point not on the boundary curve and substituting it into the inequality.
2. Use the inequality mode built into the calculator.

Example 1 Graph $3x + 4y \leq 12$

Solution: Set the RANGE to [-10,10]1 by [-10,10]1.

Keystrokes	*Screen Display*	*Explanation*
Method 1		
MODE 1 MODE +		
MODE SHIFT +		
MODE SHIFT 5		
(12 - 3 X,θ,T		Write $3x+4y=12$ as $y=(12-3x)/4$. Enter the function and store it.
) ÷ 4 SHIFT F MEM		
F1 :STO 1 AC	Graph Y=(12-3x)÷ 4	Graph. Determine the half-plane by hand by choosing the point (0,0) and substituting into the inequality. The inequality is true for this point. Hence, we want the lower half-plane.
Graph F2 :RCL 1 EXE		
Method 2		
MODE 1		Enter the inequality mode for graphing (use the division sign).
MODE +		
MODE SHIFT ÷		
MODE SHIFT 5		
SHIFT F5 :Cls EXE	Graph Y≤(12-3x)÷ 4	
AC Graph F4 :Y≤ (12		Graph using the inequality sign. The calculator automatically shades.
- 3 X,θ,T) ÷ 4 EXE		

B-17 Exponential and Hyperbolic Functions

Example 1 Graph $y = 10^{0.2x}$

Solution: Set the RANGE to [-10,10]1 by [-10,10]1 (see Section B-7 of this manual).

Keystrokes

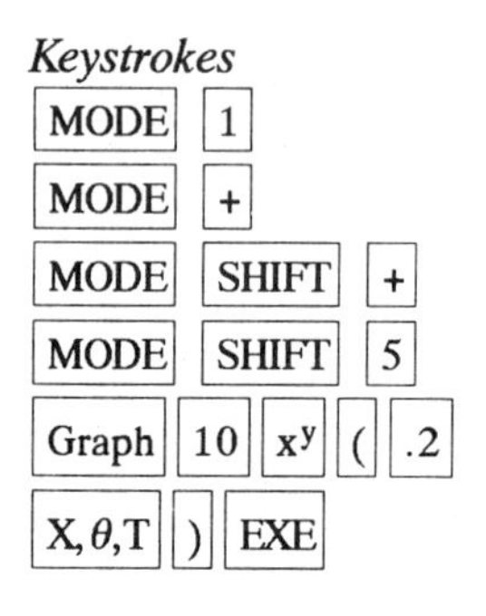

Screen Display

Graph Y=10xʸ(.2X)

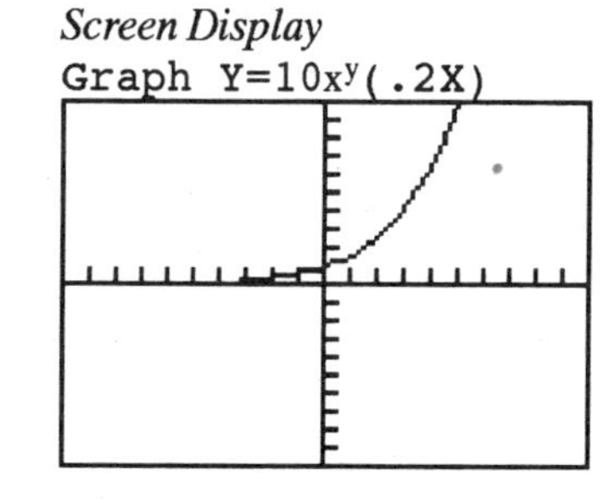

Explanation

Set the calculator to rectangular graphing mode. Enter the function and graph it.

Example 2 Graph $y = \frac{e^{x} - e^{-x}}{2}$.

Solution: Set the RANGE to [-10,10]1 by [-10,10]1 (see Section B-7 of this manual).

Keystrokes

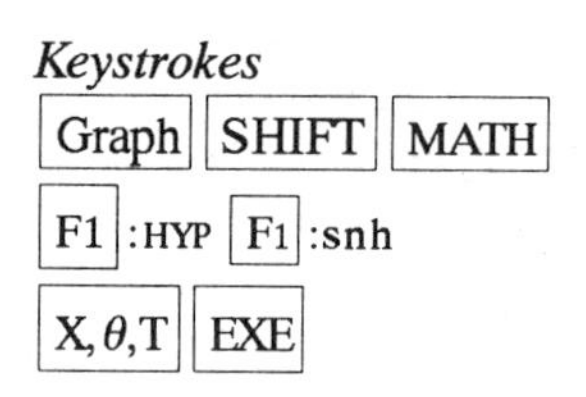

Screen Display

Graph Y=sinh X

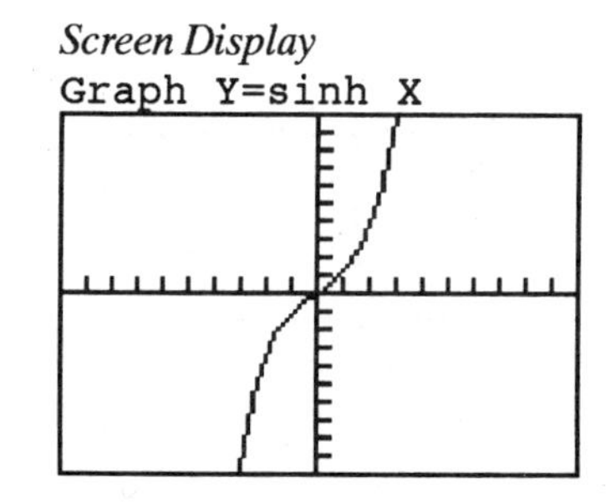

Explanation

We observe that this is the hyperbolic sine function. So we can use the built-in function in the calculator. We do not have to store the function in order to graph it.

This could also have been graphed by storing:

(SHIFT e^x X,θ,T

- SHIFT e^x SHIFT (−)

X,θ,T) ÷ 2

as a function and graphing.

B-18 Angles and Trigonometric Functions

Example 1 Evaluate to two decimal places $f(x) = \sin x$ and $g(x) = \tan^{-1}x$ at $x = \frac{5\pi}{8}$.

Solution:

Keystrokes	*Screen Display*	*Explanation*
AC		Clear the screen.
SHIFT DRG F2 :Rad EXE	Rad 0	The angle measure is given in radians. Set the calculator for radian measure before starting calculations.
sin (5 SHIFT π ÷ 8) EXE	sin (5π÷8) 0.9238795325	Enter $\sin \frac{5\pi}{8}$ and execute.
SHIFT tan⁻¹ (5 SHIFT π ÷ 8) EXE	tan⁻¹ (5π÷8) 1.099739749	Enter $\tan^{-1} \frac{5\pi}{8}$ and execute. Thus $f(\frac{5\pi}{8}) \approx 0.92$ and $g(\frac{5\pi}{8}) \approx 1.10$.

Example 2 Evaluate $f(x) = \csc x$ at $x = 32° 5' 45''$.

Solution:

Keystrokes	*Screen Display*	*Explanation*
AC		Clear the screen.
SHIFT DRG F1 :Deg EXE	Deg 0	The angle measure is given in degrees. Set the calculator for degree measure before starting calculations.
1 ÷ sin (32 + 5 ÷ 60 + 45 ÷ 3600) EXE	1÷sin (32+5÷60+4 5÷3600) 1.882044822	Enter $\csc x$ as $1 \div \sin x$. Change the minutes and seconds to decimal degrees when entering the angle measure.

Example 3 Graph $f(x) = 1.5 \sin 2x$.

Solution:

Keystrokes

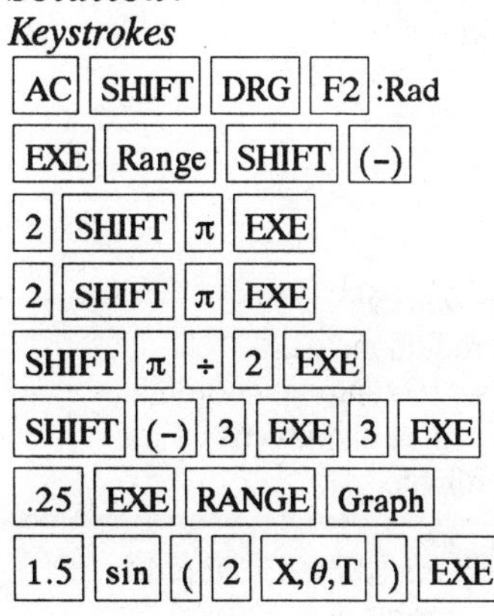

Screen Display

Rad
0.

Graph Y=1.5sin (
2X)

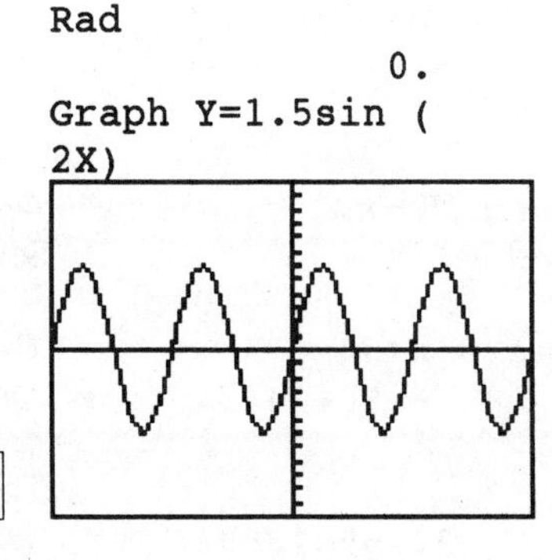

Explanation

Clear the screen.

Set the calculator for radian measure before graphing.

Set the range for $[-2\pi,2\pi]\pi/2$ by $[-3,3].25$. Note the calculator automatically changes 2π to 6.28318531 in the calculator. Press Range to see this.

Enter the function and graph.

Example 4 Graph $g(x) = 3\tan^{-1}(.2x)$.

Solution:

Keystrokes

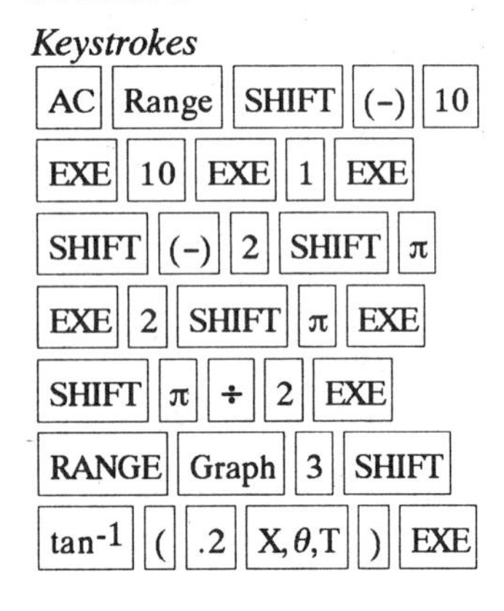

Screen Display

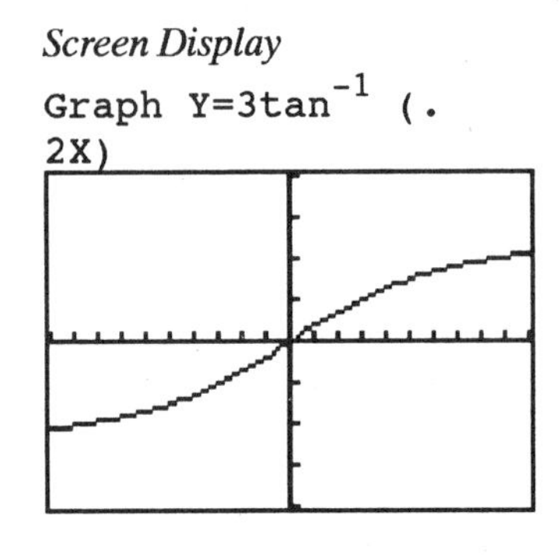

Explanation

Clear the screen.

The calculator is still in radian mode. Set the RANGE to $[-10,10]1$ by $[-2\pi,2\pi]\pi/2$

Note the calculator automatically changes 2π to 6.28318531 in the calculator. Press Range to see this.

Enter the function and graph.

B-19 Polar Coordinates

Example 1 Change $(-\sqrt{3},5)$ to polar form with $r \geq 0$ and $0 \leq \theta \leq 2\pi$.

Solution:

Keystrokes / *Screen Display*

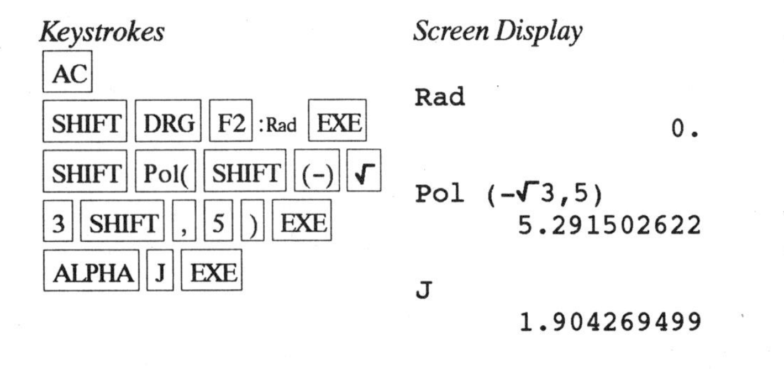

Explanation

Clear the screen.

The value displayed is *r*.

The angle measure is stored as the variable J.

Example 2 Change $(5,\pi/7)$ to rectangular coordinates.

Solution:

Keystrokes / *Screen Display*

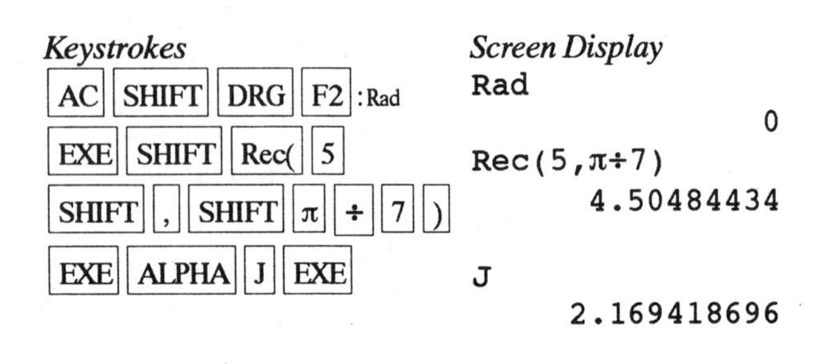

Explanation

Set radian measure.

Select rectangular coordinates.

The value displayed is the x coordinate.

The y coordinate is stored as the variable J.

Example 3 Evaluate $r = 5 - 5\sin\theta$ at $\theta = \frac{\pi}{7}$.

Solution:

Keystrokes	*Screen Display*	*Explanation*
AC		
SHIFT π ÷ 7 →	π÷7→θ .4487989505	Store $\frac{\pi}{7}$ as θ. Note that θ will be displayed since we are in polar mode.
X,θ,T EXE 5 - 5 sin		
X,θ,T EXE	5-5sin θ 2.830581304	Enter 5-5sin θ and evaluate.

Example 4 Graph $r = 5 - 5\sin\theta$

Polar equations can be graphed when in polar mode.

Solution:

Keystrokes	*Screen Display*	*Explanation*
SHIFT DRG F2 :Rad	Rad 0.	Set radian mode.
MODE SHIFT - AC		Select polar mode.
Range SHIFT (-) 10 EXE 10 EXE 1 EXE SHIFT (-) 10 EXE 10 EXE 1 EXE 0 EXE 2 SHIFT π EXE .1 EXE		Set the RANGE on x and y to [-10,10]1 by [-10,10]1. Set the range on θ to [0,2π].1.
Graph 5 - 5 sin X,θ,T EXE	Graph r=5-5sin θ	Enter the function. The graph on the standard screen is slightly distorted since the scale marks on the y axis are closer together than the scale marks on the x axis.
Range SHIFT (-) 15 EXE 15 EXE 1 EXE SHIFT (-) 10 EXE 10 EXE 1 EXE Range ◄ EXE	Graph r=5-5sin θ	Change the RANGE to [-15,15]1 by [-10,10]1 which makes the scale marks the same distance apart on both axes. The left arrow key will replay the last entry.

B-20 Scientific Notation, Significant Digits, and Fixed Number of Decimal Places

Numbers can be entered into the calculator in scientific notation.

Example 1 Calculate $(-8.513\times10^{-3})(1.58235\times10^{2})$. Enter numbers in scientific notation. Display six significant digits in the answer.

Solution:

Keystrokes	*Screen Display*	*Explanation*
[AC] [MODE] [SHIFT] [+] [AC]		Clear the screen and set rectangular mode.
[SHIFT] [DISP] [F2] :Sci [5] [EXE]	Sci 5 0.0000E+00	Set the display for 5 significant digits.
[SHIFT] [(-)] [8.513] [EXP] [SHIFT] [(-)] [3] [EXE]	⁻8.513E⁻3 ⁻8.5130E-03	Enter the first number. It is not necessary to press EXE. This is illustrated to show how the numbers are displayed.
[×] [1.58235] [EXP] [2] [EXE]	⁻8.513E⁻3×1.582 35E 2 ⁻1.3471E+00	Multiply by the second number. The answer is rounded to 5 significant digits.

Example 2 Set the scientific notation mode with six significant digits and calculate $(351.892)(5.32815\times10^{-8})$.

Solution:

Keystrokes	*Screen Display*	*Explanation*
[AC]		Clear the screen.
[SHIFT] [DISP] [F2] :Sci [6] [EXE]	Sci 6 0.00000E+00	Set the display for 6 significant digits.
[351.892] [×] [5.32815] [EXP] [SHIFT] [(-)] [8] [EXE]	351.892×5.32815E ⁻8 1.87493E-05	Enter the numbers and calculate.

Example 3 Fix the number of decimal places at 2 and calculate the interest earned on $53,218.00 in two years when invested at 5.21% simple interest.

Solution:

Keystrokes	*Screen Display*	*Explanation*
[AC]		Clear the screen.
[SHIFT] [DISP] [F1] :Fix [2] [EXE]	Fix 2 0.00	Set the display for 2 decimal places.
[53218] [×] [.0521] [×] [2] [EXE]	53218×.0521×2 5545.32	Enter the numbers and calculate. Only two decimal places are shown in the answer. The interest is $5545.32.

NOTES

APPENDIX C
TI-85 GRAPHING CALCULATOR
BASIC OPERATIONS

C-1 Getting Started

Press [ON] to turn on the calculator.

Press [CLEAR] to clear the screen.

Press [2nd] [+] to get the RESET menu. It will be displayed at the bottom of the screen. The menu is shown at the right.

RAM	DELET	RESET		

Press [F3] :RESET to get the reset menu. The first menu is now displayed in inverse shading on the line above the new menu.

RAM	DELET	RESET		
ALL	MEM	DFLTS		

Press [F1] :ALL to clear the memory.

You will get another menu with a message as shown at the right.

```
Are you sure?
```

			YES	NO

Press [F4] :YES to clear the memory.

The display now shows the message shown at the right.

Press [CLEAR] to clear the screen.

```
Mem cleared
Defaults set
```

Press [2nd] [▲] to make the display darker.

Press [2nd] [▼] to make the display lighter.

Press [2nd] [OFF] to turn off the calculator.

> A menu choice will be noted as the key to press followed by the meaning of the key.
> Example: [F3] :RESET means to press the [F3] key to choose RESET.

C-2 Calculator Operation

Home Screen

The blank screen is called the Home Screen. You can always get to this screen (aborting any calculations in progress) by pressing [2nd] [QUIT]. [QUIT] is the function above the [EXIT] key.

> Functions above the keys will be referred to as if they are keys but will be preceded by [2nd].

[2nd]

This key must be pressed to access the operation above and to the **left** of a key. An up arrow [↑] is displayed as the cursor on the screen after [2nd] key is pressed. In this manual, these functions will be referred to in square boxes just as if the function was printed on the key cap. For example [ANS] is the function above the [(-)] key.

[ALPHA]

This key must be pressed to access the operation above and to the **right** of a key. An [A] is displayed as the cursor on the screen after the [ALPHA] key is pressed.

ALPHA LOCK is engaged when the [ALPHA] key is pressed twice in succession. The calculator will remain locked in the alpha mode until the [ALPHA] key is pressed again. Alpha LOCK is useful when entering variable names that are more than one character. A variable name can be up to 8 characters in length.

[2nd] [alpha]

The key combination [2nd] [alpha] will produce lower case letters. Lower case letters are used as variables in expressions. Lower case letters are different from upper case letters.

[MODE]

Press [2nd] [MODE]. The highlighted items are active. Select the item you wish using the arrow keys. Press [ENTER] to activate the selection.

Type of notation for display of numbers.	Norm Sci Eng
Number of decimal places displayed.	Float 012345678901
Type of angle measure.	Radian Degree
Display format of complex numbers.	RectC PolarC
Function, polar, parametric, differential equation graphing.	Func Pol Param DifEq
Decimal, binary, octal or hexadecimal number base.	Dec Bin Oct Hex
Rectangular, cylindrical, or spherical vectors.	RectV CylV SphereV
Exact differentiation or numeric differentiation.	dxDer1 dxNDer

Menus

The TI-85 graphing calculator uses menus for selection of specific functions. The items on the menus are displayed across the bottom of the screen. Press the function key directly below the item on the menu you wish to choose. In this manual the menu items will be referred to using the key to be pressed followed by the meaning of the menu. For example, [F2] :RANGE refers to the second item on the [GRAPH] menu. Press [GRAPH] to see this.

[EXIT]
Press this key to exit the menu closest to the bottom of the screen.

C-3 Correcting Errors

It is easy to correct errors when entering data into the calculator by using the arrow keys, [INS], and [DEL] keys. You need to press [2nd] [INS] to insert a character before the cursor position.

[◄] or [►]	Moves the cursor to the left or right one position.
[▲]	Moves the cursor up one line.
[▼]	Moves the cursor down one line.
[DEL]	Deletes one character at the cursor position.
[2nd] [INS]	Inserts one or more characters at the cursor position.
[2nd] [ENTRY]	Replays the last executed line of input.

C-4 Calculation

Example 1 Calculate $-8 + 9^2 - \left| \frac{3}{\sqrt{2}} - 5 \right|$.

Numbers and characters are entered in the same order as you would read an expression. Do not press [ENTER] unless specifically instructed to do so in these examples. Keystrokes are written in a column but you should enter all the keystrokes without pressing the ENTER key until [ENTER] is displayed in the example.

Solution:

Keystrokes	*Screen Display*	*Explanation*
[2nd] [QUIT] [CLEAR]		It is a good idea to clear the screen before starting a calculation.
[(-)] [8] [+] [9] [^] [2] [-] [2nd] [MATH] [F1] :NUM [F5] :abs [(] [3] [÷] [2nd] [√] [2] [-] [5] [)] [ENTER]	⁻8+9^2-abs (3/√2-5) 70.1213203436	Enter numbers as you read the expression from left to right.

C-5 Evaluation of an Algebraic Expression

Example 1 Evaluate $\frac{x^4-3a}{8w}$ for $x = \pi$, $a = \sqrt{3}$, and $w = 4!$.

Two different methods can be used:

1. Store the values of the variables and then enter the expression. When [ENTER] is pressed the expression is evaluated for the stored values of the variables.

2. Store the expression and store the values of the variables. Recall the expression. Press [ENTER]. The expression is evaluated for the stored values of the variables.

The advantage of the second method is that the expression can be easily evaluated for several different values of the variables.

Solution:

Keystrokes	*Screen Display*
Method 1	
[2nd] [QUIT] [CLEAR]	
[2nd] [π] [STO▸] [x-VAR] [ENTER]	π→X
	3.14159265359
[2nd] [√] [3] [STO▸] [A] [ENTER]	√3→A
	1.73205080757
[4] [2nd] [MATH] [F2] :PROB [F1] :! [STO▸] [W] [ENTER]	4!→ W
	24

In this manual the notation [F1] :! refers to the menu item listed on the screen above the [F1] key.

[(] [ALPHA] [x-VAR] [^] [4] [-] [3] [ALPHA] [A] [)] [÷]	(X^4-3A)/(8W)
[(] [8] [ALPHA] [W] [)] [ENTER]	.480275721934

[ALPHA] is needed before [x-VAR] to get the upper case X that was used in storing the variable value. The calculator treats X and x as separate variables.

Note that [STO▸] puts the calculator in ALPHA mode automatically.

Method 2	
[2nd] [QUIT] [CLEAR]	
[GRAPH] [F1] :y(x)= [CLEAR]	
[(] [ALPHA] [x-VAR] [^] [4] [-] [3] [ALPHA] [A] [)]	y1=(X^4-3A)/(8W)
[÷] [(] [8] [ALPHA] [W] [)] [2nd] [QUIT]	

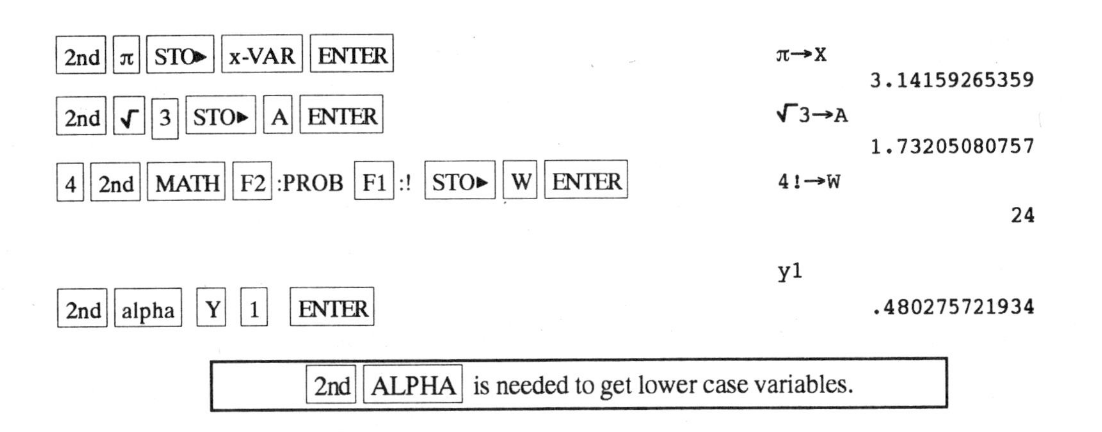

Example 2 For $f(x) = 3x+5$ and $g(x) = \sqrt{x-\sqrt{x}}$ find $f(2) - g(2)$.

Solution: (Using Method 2 of Example 1 above.)

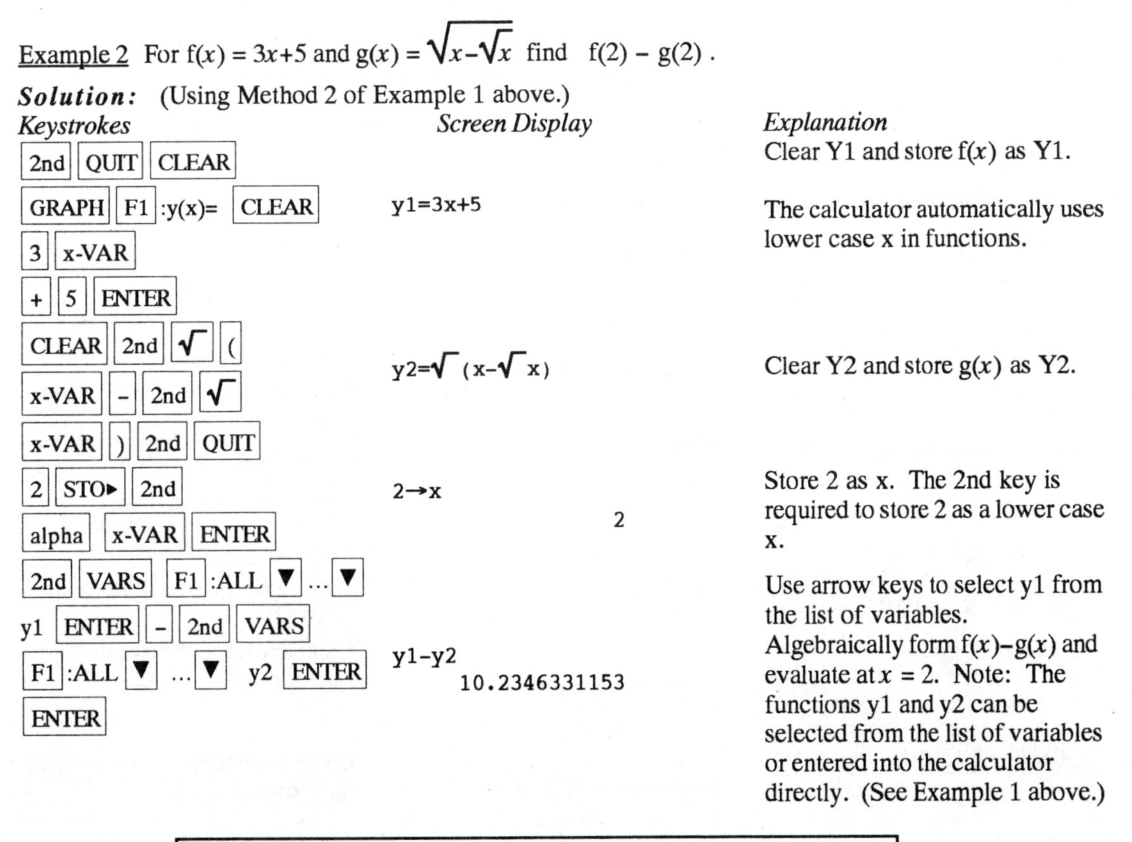

2nd alpha is needed before x-VAR to store 2 as a lower case x.

C-6 Testing Inequalities in One Variable

Example 1 Determine whether or not $x^3 + 5 < 3x^4 - x$ is true for $x = -\sqrt{2}$.

Solution:

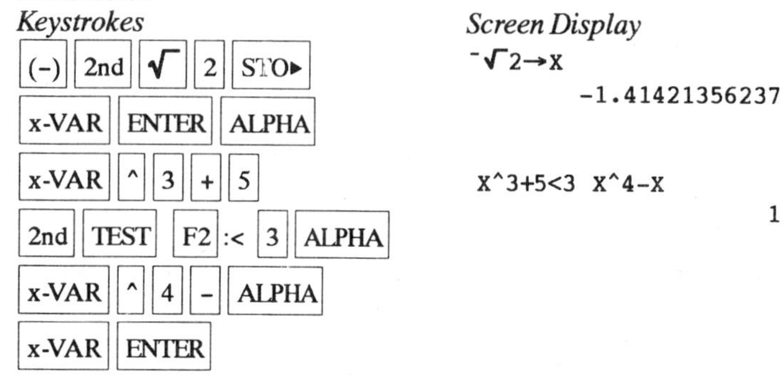

Keystrokes	*Screen Display*	*Explanation*
[(-)] [2nd] [√] [2] [STO▸]	⁻√2→x -1.41421356237	Store the value for X.
[x-VAR] [ENTER] [ALPHA]		
[x-VAR] [^] [3] [+] [5]	X^3+5<3 X^4-X	Enter the expression.
[2nd] [TEST] [F2] :< [3] [ALPHA]	1	The result of 1 indicates that the expression is true for this value of x. If a 0 was displayed, the expression would be false. The expression could have been stored as y1 and then evaluated as in Method 2 of Example 2 of Section C-5 of this manual.
[x-VAR] [^] [4] [-] [ALPHA]		
[x-VAR] [ENTER]		

> [ALPHA] is needed before [x-VAR] to get the upper case X used in expressions.

C-7 Graphing and the Standard Graphing Screen

Up to 99 functions can be stored and graphed on the same coordinate axes.

Example 1 Graph $y = x^2$, $y = .5x^2$, $y = 2x^2$, and $y = -1.5x^2$ on the same coordinate axes.

Solution:

Keystrokes	*Screen Display*	*Explanation*
[GRAPH] [F1] :y(x)= [CLEAR]		Clear the existing function and store the first function as y1.
[F1] :x [x²] [ENTER]	y1=x2	Lower case x is used in functions.
[CLEAR] [.5]	y2=.5x2	Clear and store the second function as y2.
[x-VAR] [x2] [ENTER]		
[CLEAR] [2] [x-VAR] [x2]	y3=2x2	Clear and store the third function as y3.
[ENTER] [CLEAR] [(-)]	y4=⁻1.5x2	Clear and store the fourth function as y4.
[1.5] [x-VAR] [x2] [ENTER]		
[EXIT] [F3] :ZOOM [F4] :ZSTD		Choose the Standard option from the [ZOOM] menu.

> The Standard screen automatically sets the graph for $-10 < x < 10$ and $-10 < y < 10$. Press [F2] :RANGE to see this.

The graphs will be plotted in order: y1, then y2, then y3, etc.

Occasionally you will see a vertical bar of moving dots in the upper right corner. This means the calculator is working. Wait until the dots have stopped before continuing.

There is another method that can be used to graph several functions where a coefficient or constant term has several values. This method uses the LIST feature of the calculator.

Example 2 Repeat Example 1 using LIST.

Solution:

Keystrokes	*Screen Display*	*Explanation*
GRAPH F1 :y(x)= CLEAR ... 2nd LIST F1 :{ 1 , .5 , 2 , (-) 1.5 F2 :} EXIT F1 :x x^2 ENTER	`y1={1,.5,2,-1.5}x`^{2}	Clear all of the existing functions and store the function as y1 using the LIST feature of the calculator. Lower case x is used in functions.
EXIT F3 :ZOOM F4 :ZSTD		Choose the Standard option from the ZOOM menu.

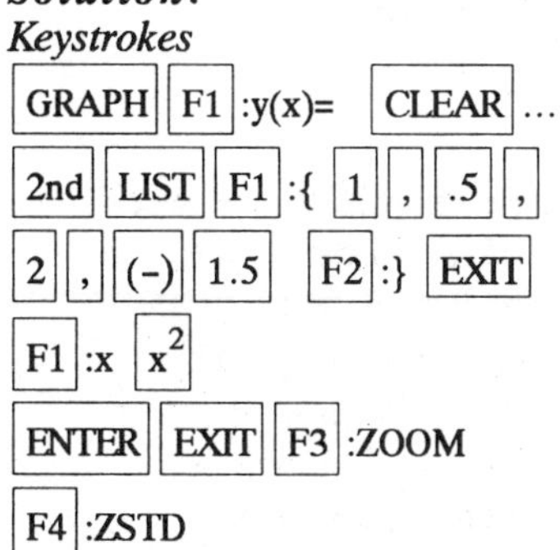

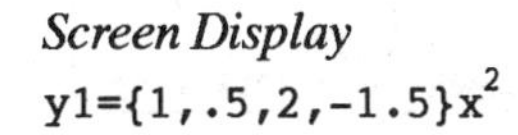

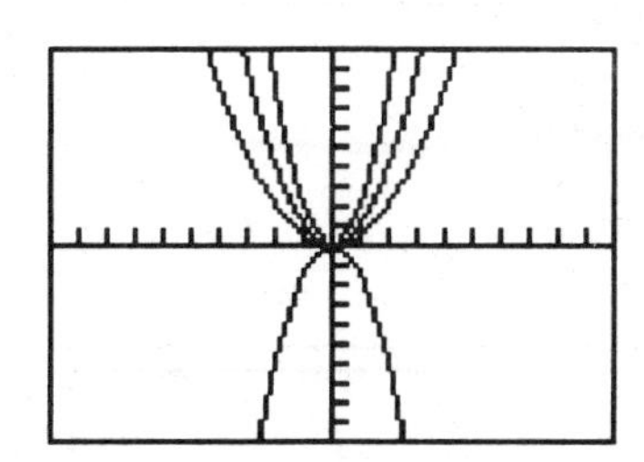

C-8 TRACE, ZOOM and RANGE

TRACE allows you to observe both the x and y coordinate of a point on the graph as the cursor moves along the graph. If there is more than one function graphed the up ▲ and down ▼ arrow keys allow you to move between the graphs displayed.

There are three methods to zoom in:

1. Change the RANGE values.
2. Set zoom factors using F1 :ZFACT on the F3 :ZOOM menu from GRAPH.
   Then use the F2 :ZIN option on the F3 :ZOOM menu from GRAPH.
3. Use the F1 :BOX option on the F3 :ZOOM menu from GRAPH.

ZOUT means to zoom out. This allows you to see a "bigger picture." (See Example 1 Section C-9 of this manual.)

ZIN means to zoom in. This will magnify a graph so the coordinates of a point can be approximated with greater accuracy.

Example 1 Approximate the value of x to two decimal places if $y = -1.58$ for $y = x^3 - 2x^2 + \sqrt{x} - 8$.

Solution:

Method 1 Change the RANGE values.

Graph the function using the Standard Graphing Screen (See Section C-7 of this manual).

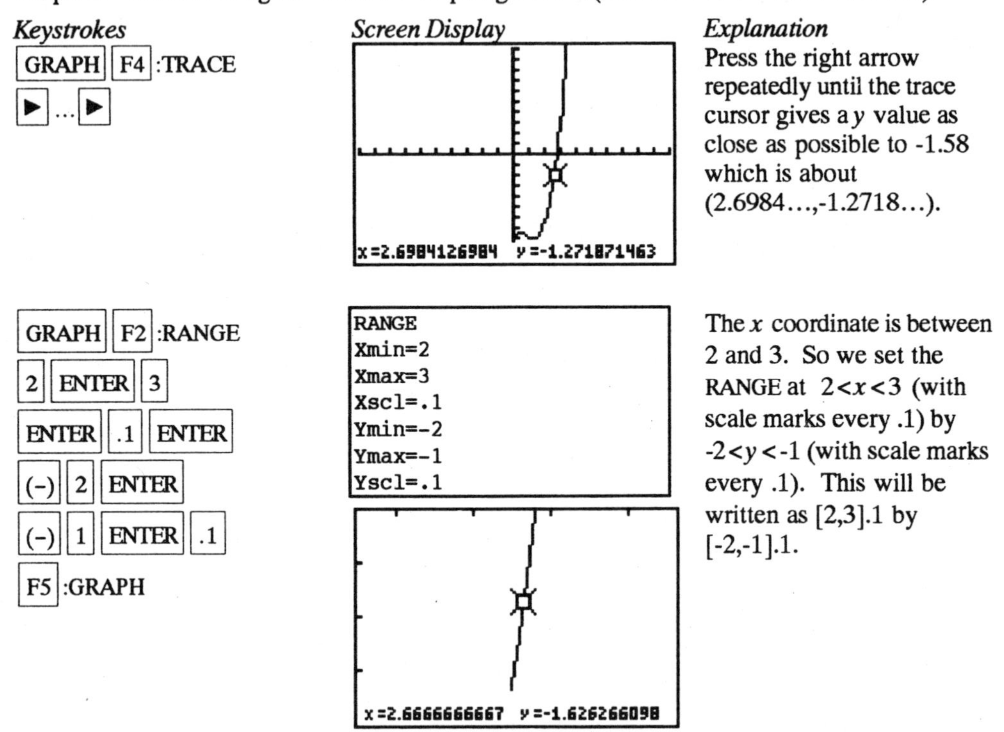

Keystrokes	*Screen Display*	*Explanation*
GRAPH F4 :TRACE ▶ ... ▶	x=2.6984126984 y=-1.271871463	Press the right arrow repeatedly until the trace cursor gives a y value as close as possible to -1.58 which is about (2.6984...,-1.2718...).
GRAPH F2 :RANGE 2 ENTER 3 ENTER .1 ENTER (-) 2 ENTER (-) 1 ENTER .1 F5 :GRAPH	RANGE Xmin=2 Xmax=3 Xscl=.1 Ymin=-2 Ymax=-1 Yscl=.1 x=2.6666666667 y=-1.626266098	The x coordinate is between 2 and 3. So we set the RANGE at $2<x<3$ (with scale marks every .1) by $-2<y<-1$ (with scale marks every .1). This will be written as [2,3].1 by [-2,-1].1.

F4 :TRACE can be used again to estimate a new x value. Repeat using TRACE and changing the RANGE until the approximation of (2.67,-1.58) has been found. Note that you may need to press the arrow keys repeatedly before the cursor becomes visible.

Method 2 Use the F2 :ZIN option on the ZOOM menu.

Graph the function using the Standard Graphing Screen (See Section C-7 of this manual).

Keystrokes	*Screen Display*	*Explanation*
GRAPH F3 :ZOOM MORE MORE	y(x)= RANGE ZOOM TRACE GRAPH ZFACT ZOOMX ZOOMY ZINT ZSTO	Get the ZOOM option from the GRAPH menu. There is a small right arrow on the screen at the right of the ZOOM menu options. This means there are more options. Press MORE twice until ZFACT option is visible.

F1 :ZFACT

5 ENTER 5

GRAPH

F3 :ZOOM F2 :ZIN

▶ ... ▼ ENTER

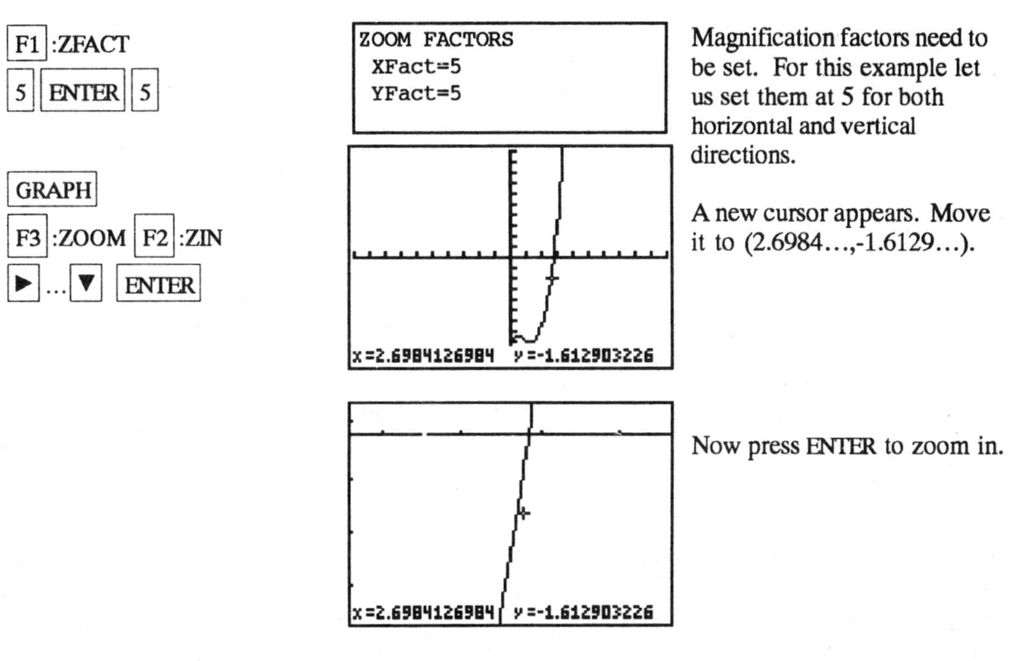

Magnification factors need to be set. For this example let us set them at 5 for both horizontal and vertical directions.

A new cursor appears. Move it to (2.6984…,-1.6129…).

Now press ENTER to zoom in.

Use trace to get a new approximation for the coordinates of the point.
Repeat this procedure until you get a value for the x coordinate accurate to two decimal places. The point has coordinates (2.67,-1.58).

<u>Method 3</u> Use the F1 :BOX option on the ZOOM menu.

Graph the function using the Standard Graphing Screen (See Section C-7 of this manual).

Keystrokes	*Screen Display*	*Explanation*
ZOOM F4 :ZSTD F1 :BOX ▶ ... ▼ ENTER ▼ ... ▶ ENTER	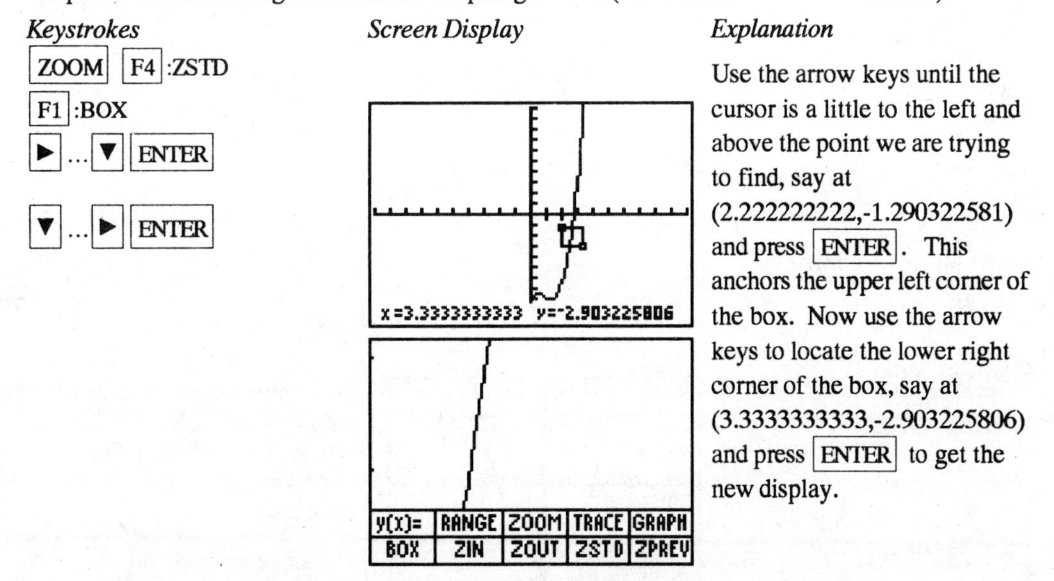	Use the arrow keys until the cursor is a little to the left and above the point we are trying to find, say at (2.222222222,-1.290322581) and press ENTER. This anchors the upper left corner of the box. Now use the arrow keys to locate the lower right corner of the box, say at (3.3333333333,-2.903225806) and press ENTER to get the new display.

Now use TRACE to approximate the coordinates of the point. Repeat this procedure until you get a value for the x coordinate accurate to two decimal places. The point has coordinates (2.67,-1.58).

C-9 Determining the RANGE

There are several ways to determine the RANGE values that should be used for the limits of the x and y axes for the screen display. Three are described below:

1. Graph using the ZSTD setting of the calculator and zoom out. The disadvantage of this method is that often the function cannot be seen at either the standard settings of [-10,10]1 by [-10,10]1 or the zoomed out settings of the RANGE.
2. Evaluate the function for several values of x. Make a first estimate based on these values.
3. Analyze the leading coefficient and the constant terms.

A good number to use for the scale marks is one that yields about 20 marks across the axis. For example if the RANGE is [-30,30] for the x axis a good scale value is (30-(-30))/20 or 3.

<u>Example 1</u> Graph the function $f(x) = .2x^2 + \sqrt[3]{x} - 32$.

Solution:
<u>Method 1</u> Use the default setting and zoom out.

Keystrokes	*Screen Display*	*Explanation*
[GRAPH] [y(x)=] [CLEAR] ...	y1=.2 x^2+x^(1/3)-32	Clear all functions. Then enter the function.
[.2] [x-VAR] [^]		
[2] [+] [x-VAR] [^] [(]		
[1] [÷] [3] [)]		
[-] [32] [ENTER]		
[EXIT] [F3] :ZOOM [F4] :ZSTD		Graph using the standard screen. Nothing is seen on the graph screen because no part of this curve is in this RANGE.
		Set the zoom factors to 5 and 5.
[F3] :ZOOM		
[MORE] [MORE] [F1] :ZFACT		
[5] [ENTER] [5]		
[F3] :ZOOM	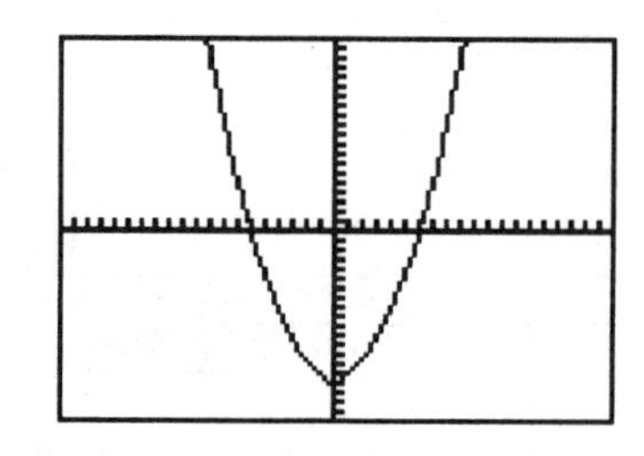	Zooming out shows a parabolic shaped curve. Note the double axis. This indicates that the scale marks are very close together.
[F3] :ZOUT [ENTER]		

Method 2 Evaluate the function for several values of x to one decimal place accuracy. (See Section C-5 of this manual on how to evaluate a function at given values of x.) Be sure to use [STO▶] [ALPHA] [x-VAR] to get lower case x variables on the calculator.

x	f(x)
-20	45.3
-10	-14.2
0	-32.0
10	-9.8
20	-50.7

Analyzing this table indicates that a good RANGE to start with is [−20,20]2 by [−50,50]5. Note the scale is chosen so that about 20 scale marks will be displayed along each of the axes.

[GRAPH] [F2] :RANGE [(-)] [20] [ENTER] [20] [ENTER] [(-)] [50] [ENTER] [50] [ENTER] [F5] :GRAPH.

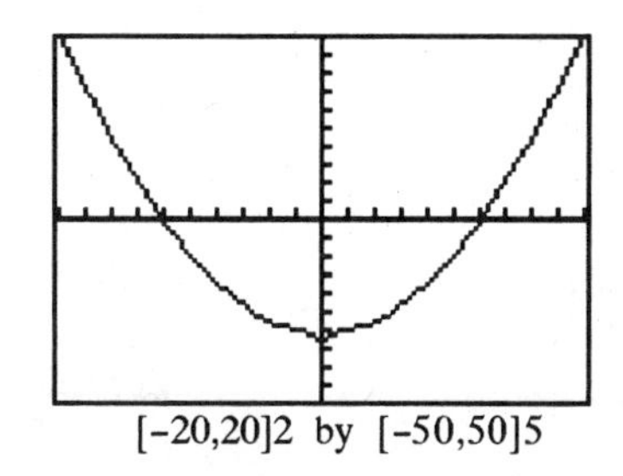

[−20,20]2 by [−50,50]5

Method 3 Analyze the leading coefficient and constant terms. Since the leading coefficient is .2 the first term will increase 2 units for each 10 units x^2 increases. This is about $\sqrt{10}$ or about 3 units increase in x. A first choice for the x-axis limits can be found using:

$$\frac{10\times(\text{unit increase in } x)}{(\text{first term increase})} = \frac{10\times 3}{2} = 15.$$ So set Xmin = -15 and Xmax = 15.

A first choice for the scale on the x axis (having about 20 marks on the axis) can be found using $\frac{\text{Xmax-Xmin}}{20} = \frac{15-(-15)}{20} = 1.5$ (round to 2). So the limits on the x axis could be [-15,15]2.

A first choice for the y-axis limits could be ±(constant term). The scale for the y axis can be found using $\frac{\text{Ymax-Ymin}}{20} = \frac{32-(-32)}{20} = 3.2$ (round to 4). So a first choice for the y-axis limits could be [-32,32]4. Hence a good first setting for the the RANGE if [-15,15]2 by [-32,32]4.

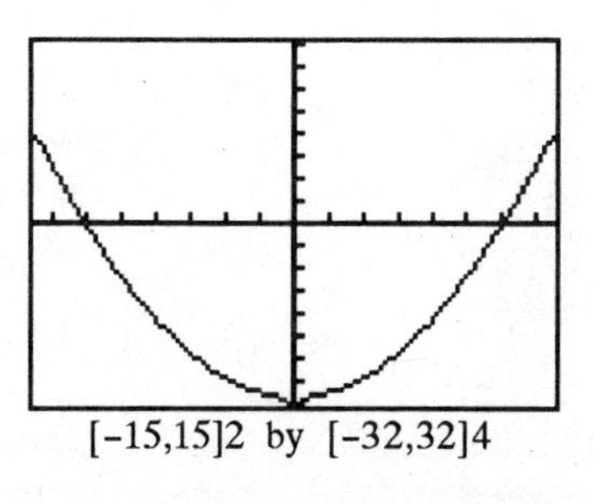

[−15,15]2 by [−32,32]4

> A good choice for the **scale** is so that about 20 marks appear along the axis. This is $\frac{\text{Xmax-Xmin}}{20}$ (rounded up to the next integer) for the x axis and $\frac{\text{Ymax-Ymin}}{20}$ (rounded up to the next integer) for the y axis.

C-10 Piecewise-Defined Functions

There are two methods to graph piecewise-defined functions:

1. Graph each piece of the function separately as an entire function on the same coordinate axes. Use trace and zoom to locate the partition value on each of the graphs.
2. Store each piece of the function separately using the dotted mode but include an inequality statement following the expression which will set the RANGE of values on x for which the function should be graphed. Then graph all pieces on the same coordinate axes.

Example 1 Graph $f(x) = \begin{cases} x^2+1 & x < 1 \\ 3x-5 & x \geq 1 \end{cases}$

Solution:

Method 1

Keystrokes	*Screen Display*	*Explanation*
GRAPH y(x)= CLEAR x-VAR ^ 2 + 1 ENTER 3 x-VAR - 5 EXIT F3 :ZOOM F4 :ZSTD EXIT F4 :TRACE ▶ ... ▶	y1=x^2+1 y2=3 x-5 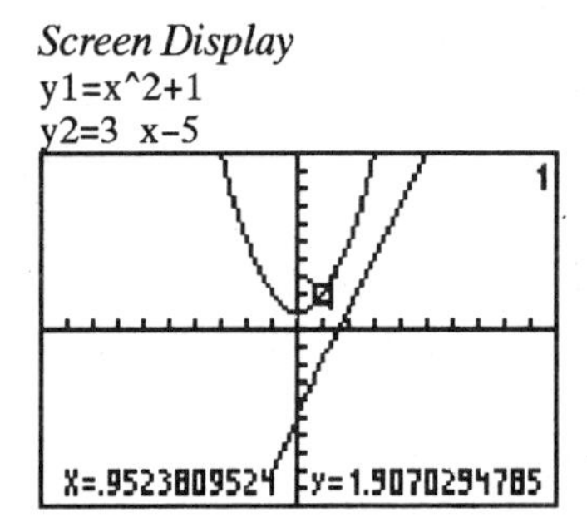	Clear all existing functions. Store the new functions. Graph. Both functions will be displayed. Use trace and zoom to find the point on the graphs where x is close to 1. The up and down arrow keys will move the cursor between the graphs. The endpoint of the parabolic piece of the graph is not included on the graph since x<1. The endpoint of the straight line piece of the graph is included. The graph shown to the left shows the curves with the cursor on the parabolic piece of the graph.

Method 2

Keystrokes	*Screen Display*	*Explanation*
GRAPH y(x)= CLEAR (x-VAR ^ 2 + 1) ÷ (x-VAR 2nd TEST F2 :< 1) ENTER	y1=(x^2+1)/(x<1)	Clear all existing functions. The logical statement $x<1$ will give a 1 when the value of x is less than 1 and a 0 when the value of x is greater than or equal to 1. Hence the first part of the function is divided by 1 when $x<1$ and 0 when $x \geq 1$. The function will not graph when it is divided by 0.

> The number of the function being traced appears in the upper right corner of the screen.

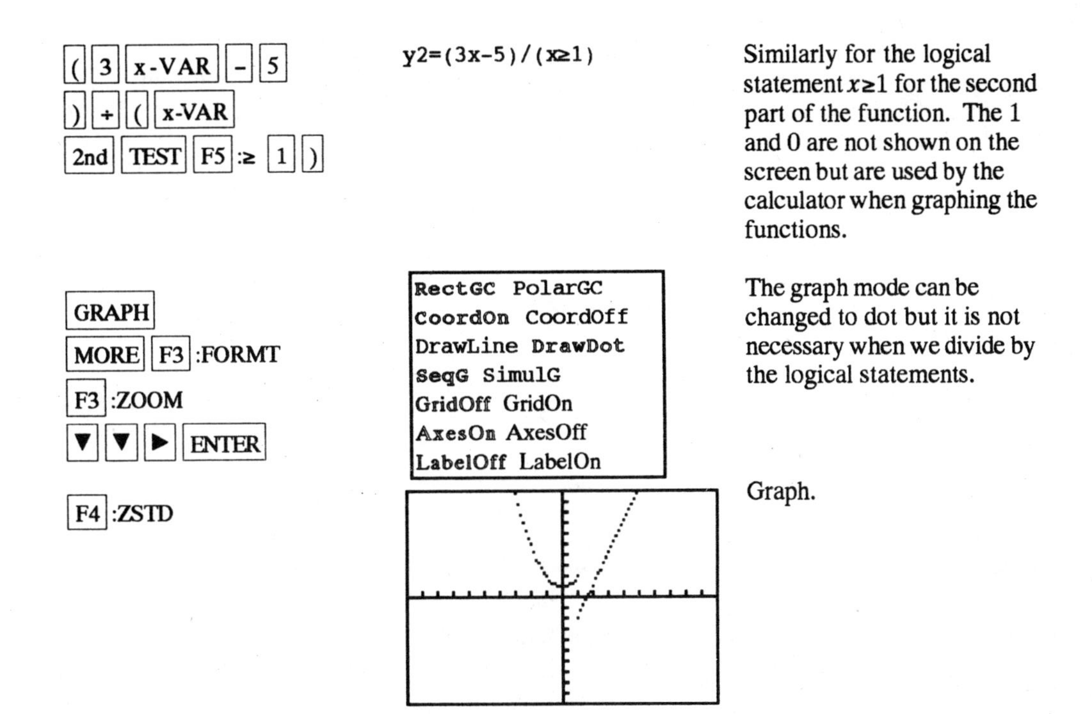

Similarly for the logical statement $x \geq 1$ for the second part of the function. The 1 and 0 are not shown on the screen but are used by the calculator when graphing the functions.

The graph mode can be changed to dot but it is not necessary when we divide by the logical statements.

Graph.

C-11 Solving Equations in One Variable

There are three methods for approximating the solution of an equation using graphing.

1. Write the equation as an expression equal to zero. Graph y=(the expression). Find where the curve crosses the x axis. The x values (x intercepts) are the solutions to the equation.

2. Graph y = (left side of the equation) and y=(right side of the equation) on the same coordinate axes. The x coordinate of the points of intersection are the solutions to the equation.

3. Use the SOLVER function of the calculator.

Example 1 Solve $\frac{3x^2}{2} - 5 = \frac{2(x+3)}{3}$.

Solution:

Method 1 Write the equation as $\left(\frac{3x^2}{2} - 5\right) - \left(\frac{2(x+3)}{3}\right) = 0$. Graph $y = \left(\frac{3x^2}{2} - 5\right) - \left(\frac{2(x+3)}{3}\right)$ and find the x value where the graph crosses the x axis. This is the x intercept.

Keystrokes

GRAPH F1 :y(x)=
CLEAR (3
x-VAR ^ 2 ÷ 2
- 5) - (2
(x-VAR + 3)
÷ 3) EXIT
F3 :ZOOM F4 :ZSTD

Screen Display

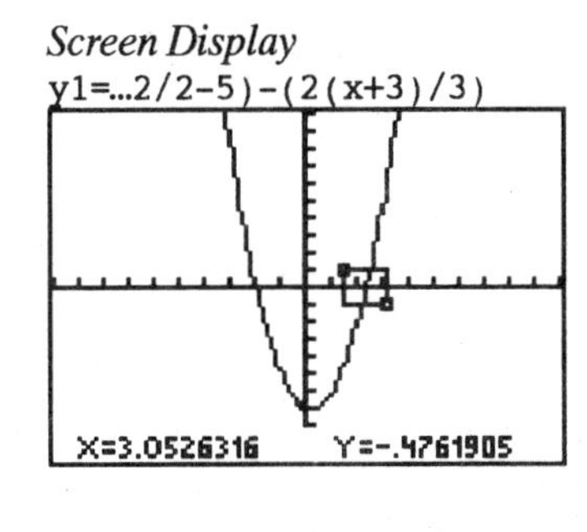

Explanation

Store the expression as Y1. The ... means there is some of the expression not shown on the display. Use the arrow keys to see the rest of the expression.

Use zoom box to find the x intercepts. A typical zoom box is shown on the graph at the left.

The solutions are: $x \approx -1.95$ and $x \approx 2.39$.

Method 2 Graph $y = \frac{3x^2}{2} - 5$ and $y = \frac{2(x+3)}{3}$ on the same coordinate axes and find the x coordinate of their points of intersection.

Keystrokes / *Screen Display*

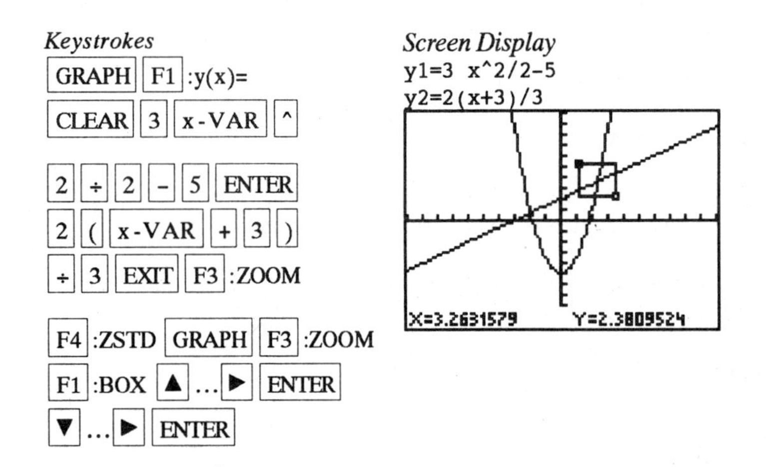

Explanation

Clear any existing functions.
Store the two functions.

Find the points of intersection. Use trace and zoom box to find the x values: $x \approx -1.95$ and $x \approx 2.39$.

A typical zoom box is shown on the graph at the left.

Hence the approximate solutions to this equation are -1.95 and 2.39.

Method 3 The calculator has the capability of solving equations. We need to enter the equation into the calculator in the SOLVER mode in order to approximate the a solution. The calculator then finds the x value closest to the approximation.

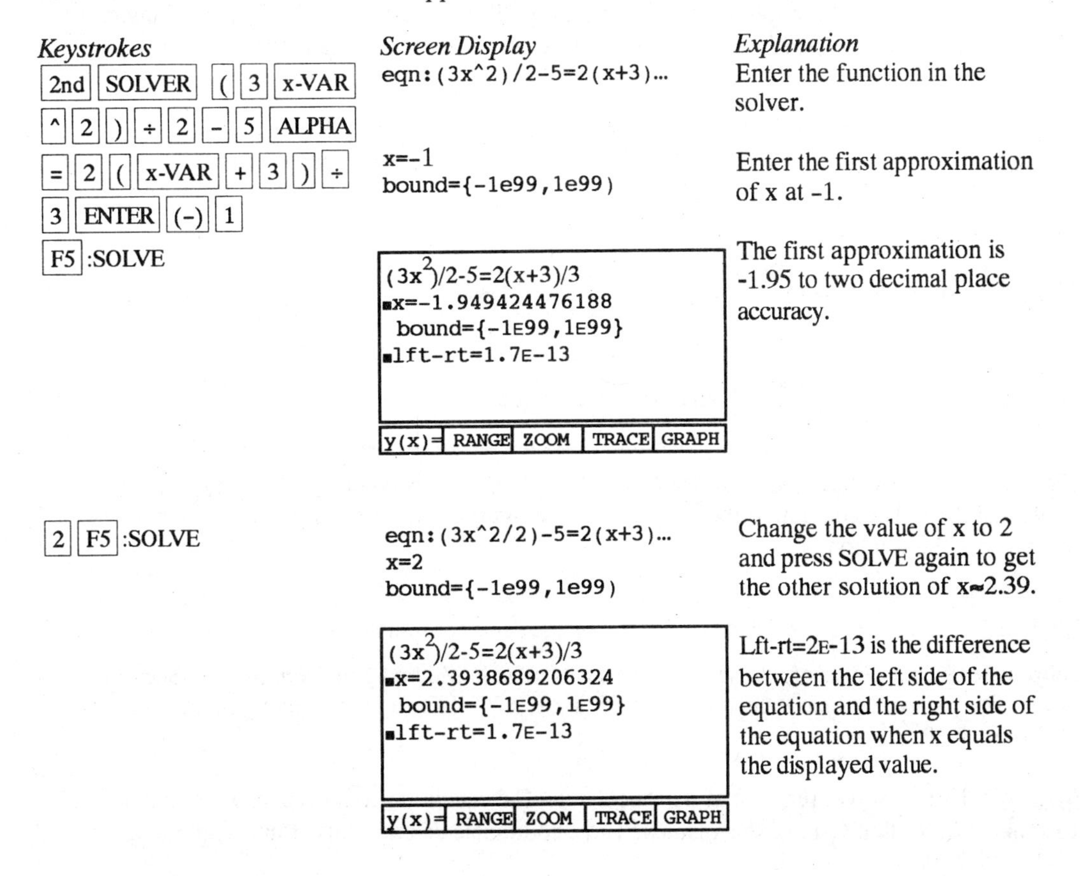

Keystrokes	*Screen Display*	*Explanation*
2nd SOLVER (3 x-VAR ^ 2) ÷ 2 - 5 ALPHA = 2 (x-VAR + 3) ÷ 3 ENTER (-) 1	eqn:(3x^2)/2-5=2(x+3)...	Enter the function in the solver.
	x=-1 bound={-1e99,1e99)	Enter the first approximation of x at -1.
F5 :SOLVE	(3x²)/2-5=2(x+3)/3 ▪x=-1.949424476188 bound={-1E99,1E99} ▪lft-rt=1.7E-13 y(x)= RANGE ZOOM TRACE GRAPH	The first approximation is -1.95 to two decimal place accuracy.
2 F5 :SOLVE	eqn:(3x^2/2)-5=2(x+3)... x=2 bound={-1e99,1e99)	Change the value of x to 2 and press SOLVE again to get the other solution of x≈2.39.
	(3x²)/2-5=2(x+3)/3 ▪x=2.3938689206324 bound={-1E99,1E99} ▪lft-rt=1.7E-13 y(x)= RANGE ZOOM TRACE GRAPH	Lft-rt=2E-13 is the difference between the left side of the equation and the right side of the equation when x equals the displayed value.

The ▪ indicates the answer found by the calculator. The ▪ disappears when you enter a value for x as a first approximation.

If you get an error message press F5 :QUIT 2nd SOLVER ENTER to begin again.

C-12 Solving Inequalities in One Variable

There are three methods for approximating the solution of an inequality using graphing.

1. Write the inequality with zero on one side of the inequality sign and the expression on the other side. Graph y=(the expression). Find the x intercepts. The solution will be an inequality with the x values (x intercepts) as the cut off numbers.

2. Graph y=(left side of the inequality) and y=(right side of the inequality) on the same coordinate axes. The x coordinate of the points of intersection are the solutions to the equation. Identify which side of the x value satisfies the inequality by observing the graphs of the two functions.

3. Use the solver to find the x value where the left side of the inequality is equal to the right side. Suppose this value is a. Choose a value of x greater than a to use as a test value. Evaluate the inequality for this test value. If the inequality is true then the interval that includes this value of x is part of the solution to the inequality. If the inequality is false then the interval that includes this value of x is not part of the solution. Repeat using a test value smaller than a.

Example 1 Approximate the solution to $\frac{3x^2}{2} - 5 \le \frac{2(x+3)}{3}$. Use two decimal places.

Solution:

Method 1 Write the equation as $\left(\frac{3x^2}{2} - 5\right) - \left(\frac{2(x+3)}{3}\right) \le 0.$

Graph $y = \left(\frac{3x^2}{2} - 5\right) - \left(\frac{2(x+3)}{3}\right)$ and find the x intercept(s).

This was done in Method 1 of Example 1 in Section C-11 of this manual. The x intercepts are -1.95 and 2.39. The solution to the inequality is the interval on x for which the graph is below the x axis. The solution is $-1.95 \le x \le 2.39$.

Method 2 Graph $y = \frac{3x^2}{2} - 5$ and $y = \frac{2(x+3)}{3}$ on the same coordinate axes and find the x coordinate of their points of intersection. This was done in Method 2 of Example 1 in Section C-11. The parabola is below the line for $-1.95 \le x \le 2.39$. Hence the inequality is satisfied for $-1.95 \le x \le 2.39$.

Method 3 Use the solver function of the calculator. The equation was solved for x in Method 3 of Example 1 in Section C-11 of this manual. The x coordinate of the points of intersections are -1.95 and 2.39.

Choose -2 as a test value. Evaluating the original inequality using the calculator yields a 0 which means the inequality is not true for this value of x. (See Section C-6 of this manual.) Repeat the testing using 0 and 3. We see that the inequality is true for $x=0$ and not true for $x=3$. Hence the inequality is satisfied for
$-1.95 \le x \le 2.39$.

C-13 Storing an Expression That Will Not Graph

Expressions can be stored as a variable. Variable names can be up to eight characters in length. The expressions can then be recalled and graphed using *y(x)=* on the graph menu.

Example 1 Store the expression B^2-4AC so that it will not be graphed but so that it can be evaluated at any time. Evaluate this expression for A=3, B=2.58, and C=$\sqrt{3}$.

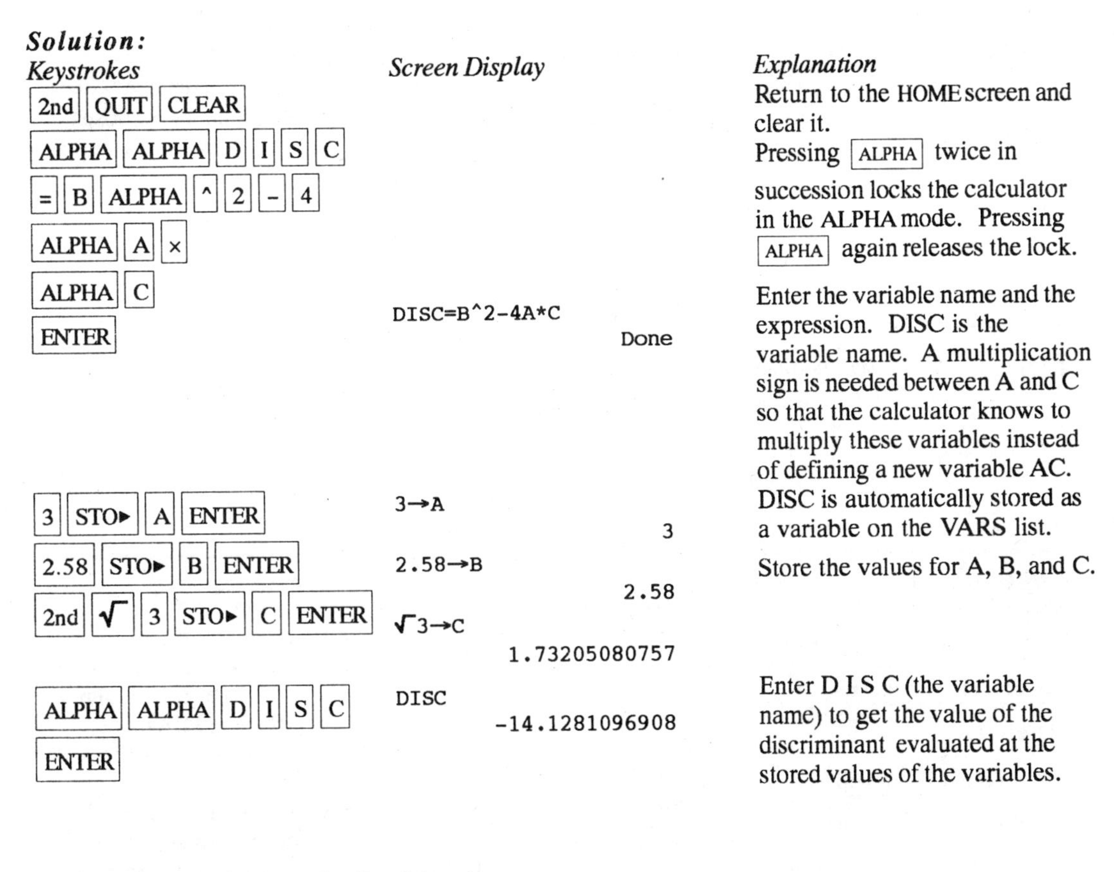

Solution:

Keystrokes	*Screen Display*	*Explanation*
[2nd] [QUIT] [CLEAR]		Return to the HOME screen and clear it.
[ALPHA] [ALPHA] [D] [I] [S] [C]		Pressing [ALPHA] twice in succession locks the calculator in the ALPHA mode. Pressing [ALPHA] again releases the lock.
[=] [B] [ALPHA] [^] [2] [-] [4]		
[ALPHA] [A] [×]		
[ALPHA] [C]		Enter the variable name and the expression. DISC is the variable name. A multiplication sign is needed between A and C so that the calculator knows to multiply these variables instead of defining a new variable AC. DISC is automatically stored as a variable on the VARS list.
[ENTER]	`DISC=B^2-4A*C` Done	
[3] [STO▸] [A] [ENTER]	`3→A` 3	Store the values for A, B, and C.
[2.58] [STO▸] [B] [ENTER]	`2.58→B` 2.58	
[2nd] [√] [3] [STO▸] [C] [ENTER]	`√3→C` 1.73205080757	
[ALPHA] [ALPHA] [D] [I] [S] [C]	`DISC` -14.1281096908	Enter D I S C (the variable name) to get the value of the discriminant evaluated at the stored values of the variables.
[ENTER]		

C-14 Permutations and Combinations

Example 1 Find (A) $P_{10,3}$ and (B) $C_{12,4}$

Solution (A):

Keystrokes

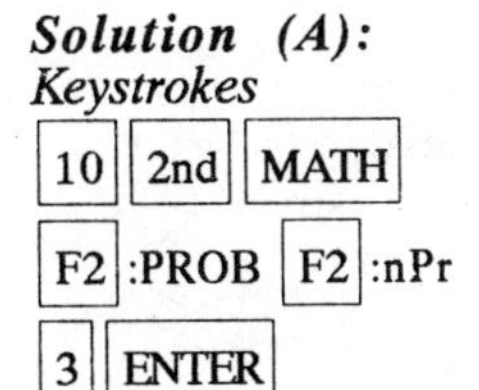

[10] [2nd] [MATH]

[F2] :PROB [F2] :nPr

[3] [ENTER]

Screen Display

10

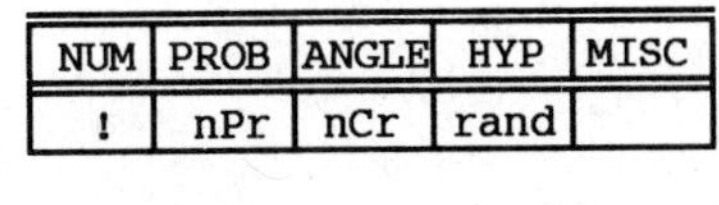

NUM	PROB	ANGLE	HYP	MISC
!	nPr	nCr	rand	

10 nPr 3

720

Explanation

Enter the first number. Get the math menu and choose PROB using the function keys. Choose `nPr`.

Solution (B):

Keystrokes	*Screen Display*	*Explanation*
12 2nd MATH	12	Enter the first number. Get the math menu and choose PROB using the arrow keys. Choose nCr.
F2 :PROB	NUM PROB ANGLE HYP MISC	
F3 :nCr 4 ENTER	! nPr nCr rand 12 nCr 4 495	

C-15 Matrices

Example 1 Given the matrices

$$A = \begin{bmatrix} 1 & -2 \\ 3 & 0 \\ 5 & -8 \end{bmatrix} \quad B = \begin{bmatrix} 2 & 1 & 5 \\ 3 & 2 & -1 \\ 0 & 8 & -3 \end{bmatrix} \quad C = \begin{bmatrix} 1 \\ -5 \\ 10 \end{bmatrix}$$

Find (A) -3BC (B) B^{-1} (C) A^T (D) det B

Solution (A):

Keystrokes	*Screen Display*	*Explanation*
2nd MATRX F2 :EDIT B ENTER 3 ENTER 3 ENTER 2 ENTER 1 ENTER 5 ENTER 3 ENTER 2 ENTER (-) 1 ENTER 0 ENTER 8 ENTER (-) 3 ENTER EXIT	MATRIX Name=B MATRIX:B 3x3 1,1=2 2,1=3 3,1=0 MATRIX:B 3x3 1,2=1 2,2=2 3,2=8 MATRIX:B 3x3 1,3=5 2,3=-1 3,3=-3 MATRIX:C 3x1 1,3=1 2,3=-5 3,3=10	Enter the matrix mode. Choose EDIT. Name the matrix B. Note the calculator is already in ALPHA mode. Set the dimensions of the matrix. Enter the elements. The calculator moves across the rows identifying the position of the element to be entered. Enter all the elements row by row. Press EXIT to exit the matrix mode. (Note: To move to the next column, press ENTER.)
F2 :EDIT C ENTER 3 ENTER 1 ENTER 1 ENTER (-) 5 ENTER 10 ENTER EXIT EXIT		Repeat this procedure to enter the elements of matrix C.

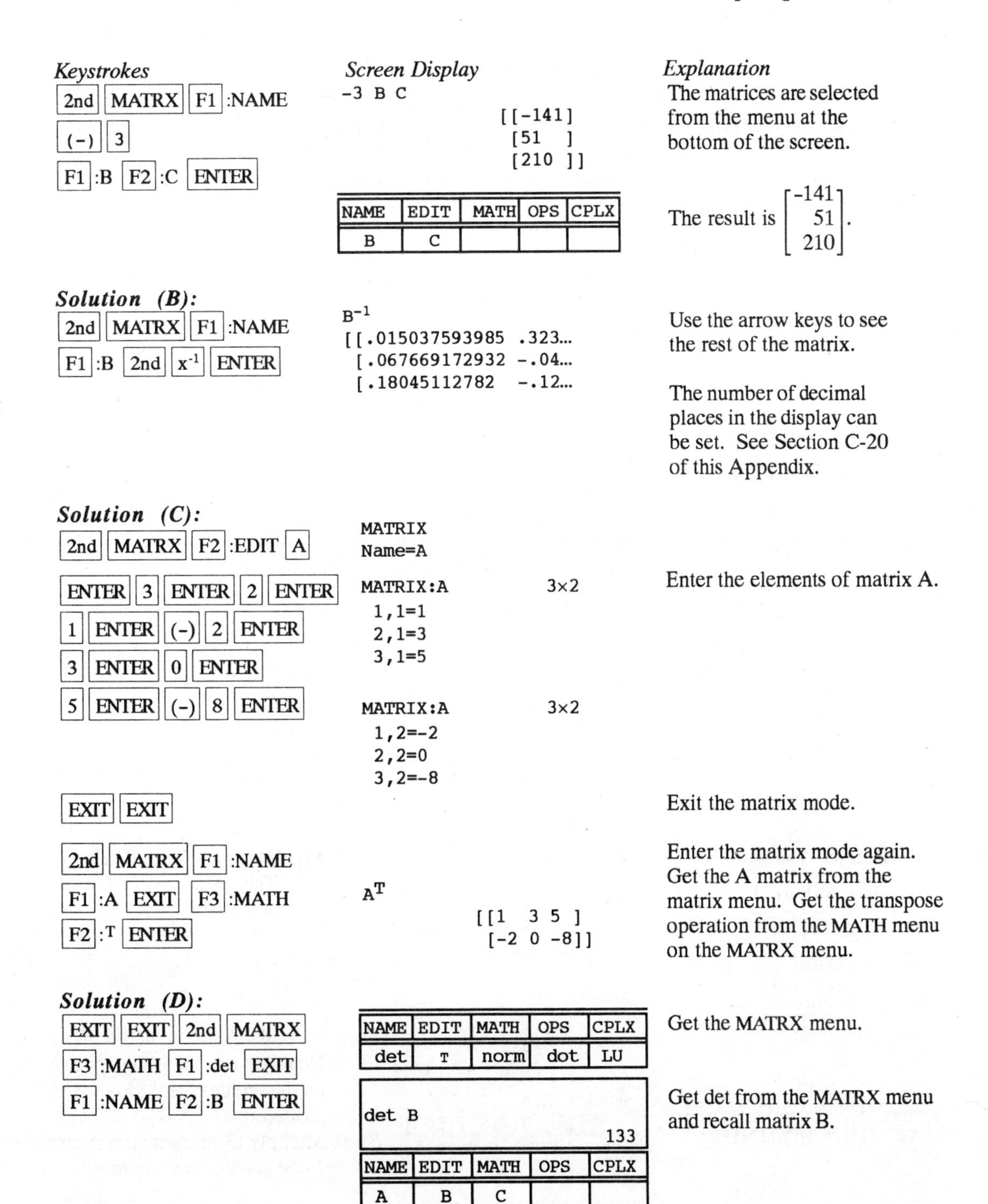

Keystrokes	*Screen Display*	*Explanation*
[2nd] [MATRX] [F1] :NAME [(-)] [3] [F1] :B [F2] :C [ENTER]	-3 B C [[-141] [51] [210]] NAME | EDIT | MATH | OPS | CPLX B | C	The matrices are selected from the menu at the bottom of the screen. The result is $\begin{bmatrix} -141 \\ 51 \\ 210 \end{bmatrix}$.
Solution (B): [2nd] [MATRX] [F1] :NAME [F1] :B [2nd] [x^{-1}] [ENTER]	B^{-1} [[.015037593985 .323... [.067669172932 -.04... [.18045112782 -.12...	Use the arrow keys to see the rest of the matrix. The number of decimal places in the display can be set. See Section C-20 of this Appendix.
Solution (C): [2nd] [MATRX] [F2] :EDIT [A]	MATRIX Name=A	
[ENTER] [3] [ENTER] [2] [ENTER] [1] [ENTER] [(-)] [2] [ENTER] [3] [ENTER] [0] [ENTER] [5] [ENTER] [(-)] [8] [ENTER]	MATRIX:A 3×2 1,1=1 2,1=3 3,1=5 MATRIX:A 3×2 1,2=-2 2,2=0 3,2=-8	Enter the elements of matrix A.
[EXIT] [EXIT]		Exit the matrix mode.
[2nd] [MATRX] [F1] :NAME [F1] :A [EXIT] [F3] :MATH [F2] :T [ENTER]	A^T [[1 3 5] [-2 0 -8]]	Enter the matrix mode again. Get the A matrix from the matrix menu. Get the transpose operation from the MATH menu on the MATRX menu.
Solution (D): [EXIT] [EXIT] [2nd] [MATRX]	NAME | EDIT | MATH | OPS | CPLX det | T | norm | dot | LU	Get the MATRX menu.
[F3] :MATH [F1] :det [EXIT] [F1] :NAME [F2] :B [ENTER]	det B 133 NAME | EDIT | MATH | OPS | CPLX A | B | C	Get det from the MATRX menu and recall matrix B.

Example 2 Find the reduced form of matrix $\begin{bmatrix} 2 & 1 & 5 & 1 \\ 3 & 2 & -1 & -5 \\ 0 & 8 & -3 & 10 \end{bmatrix}$.

Delete existing matrices. Enter the dimensions and elements of the 3x4 matrix of this problem using the procedure in Example 1 of Section C-15 above.

Solution:

Keystrokes	*Screen Display*	*Explanation*
2nd MEM F2 :DELET MORE	MATRIX Name=A	Delete the existing matrices.
F1 :MATRX ENTER ENTER ENTER EXIT 2nd MATRX F2 :EDIT	MATRIX:A 3x4 1,1=2 2,1=3 3,1=0	Enter the matrix mode. Enter the dimensions and the elements.
A ENTER 3 ENTER 4 ENTER 2 ENTER 1 ENTER 5 ENTER 1 ENTER	MATRIX:A 3x4 1,2=1 2,2=2 3,2=8	
3 ENTER 2 ENTER (-) 1 ENTER (-) 5 ENTER 0 ENTER 8 ENTER	MATRIX:A 3x4 1,3=5 2,3=-1 3,3=-3	
(-) 3 ENTER 10 ENTER	MATRIX:A 3x4 1,4=1 2,4=-5 3,4=10	
EXIT EXIT 2nd MATRX F4 :OPS MORE F4 :multR .5 , ALPHA A , 1) ENTER	multR(.5,A,1) [[1 .5 2.5 .5] [3 2 -1 -5] [0 8 -3 10]]	Multiply row 1 of matrix A by .5. The result is stored in the temporary memory ANS.
STO▸ ALPHA A ENTER	Ans→A [[1 .5 2.5 .5] [3 2 -1 -5] [0 8 -3 10]]	Store the result in matrix A. Note Ans automatically appears on the screen when STO▸ is pressed.
F5 :mRAdd (-) 3 , ALPHA A , 1 , 2) ENTER	mRAdd(-3,A,1,2) [[1 .5 2.5 .5] [0 .5 -8.5 -6.5] [0 8 -3 10]]	Multiply -3 times matrix A row 1 and add the result to row 2.

Keystrokes	Screen Display	Explanation

[STO▸] [A] [ENTER]	`Ans→A`
	`[[1 .5 2.5  .5  ]`
	`[0 .5 -8.5 -6.5]`
	`[0 8  -3   10  ]]`
		Store the result in matrix A.

[F4] :multR [2] [,] [ALPHA] [A] [,] [2] [)] [ENTER]	`multR(2,A,2)`
	`[[1 .5 2.5  .5 ]`
	`[0 1  -17  -13]`
	`[0 8  -3   10 ]]`
		2 times matrix A row 2.

[STO▸] [A]	`Ans→A`
	`[[1 .5 2.5  .5 ]`
	`[0 1  -17  -13]`
	`[0 8  -3   10 ]]`
		Store the result in matrix A.

Continue using row operations to arrive at the reduced form of $\begin{bmatrix} 1 & 0 & 0 & -2.428... \\ 0 & 1 & 0 & 1.571... \\ 0 & 0 & 1 & .857... \end{bmatrix}$

C-16 Graphing an Inequality

There are two ways to graph an inequality.

1. Graph the boundary curve. Determine the half-plane by choosing a test point not on the boundary curve and substituting the test value into the inequality.

2. Repeat Method 1 to determine which side of the graph is to be shaded. Use the SHADE option on the calculator to get a shaded graph.

Example 1 Graph $3x + 4y \leq 12$.

Solution:

Keystrokes	*Screen Display*	*Explanation*
Method 1 [GRAPH] [F1] :y(x)= [CLEAR] [(] [12] [-] [3] [x-VAR] [)] [÷] [4] [EXIT] [F3] :ZOOM [F4] :ZSTD	`y1=(12-3 x)/4`	Graph $3x+4y=12$ by first writing as $y=(12-3x)/4$. Determine the half-plane by choosing the point (0,0) and substituting into the inequality by hand. The inequality is true for this point. Hence, we want the lower half-plane.

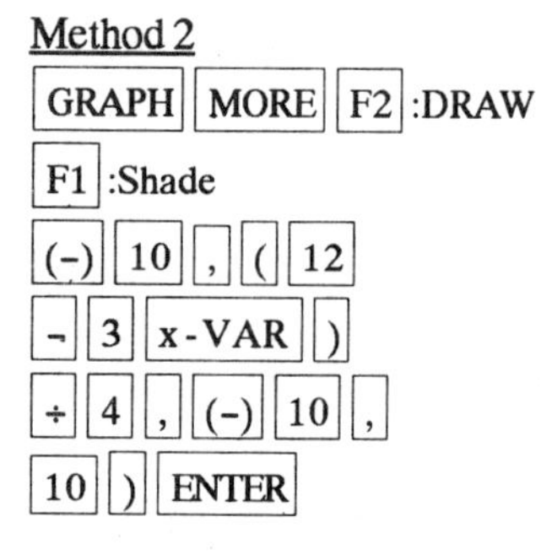

`Shade(-10,(12-3 x)/4,`
`-10,10)`

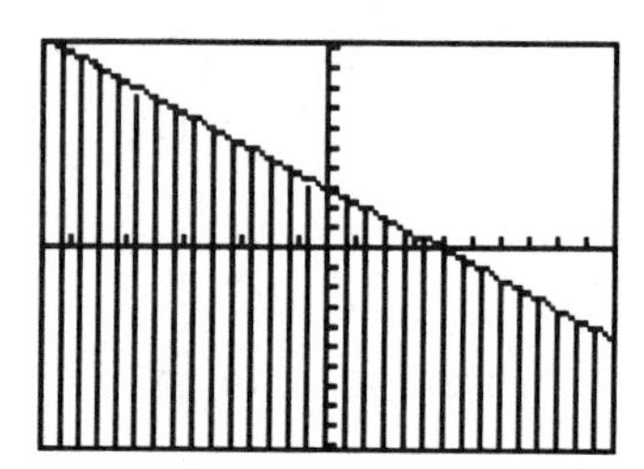

Repeat Method 1 to determine we want the lower half-plane.

Enter the Shade command. The numbers in the Shade command are

Lower boundary	(a function)
Upper boundary	(a function)
Left boundary	(a number)
Right boundary	(a number)

REMINDER: Commas are needed between entries in the shade command.

C-17 Exponential and Hyperbolic Functions

Example 1 Graph $y = 10^{0.2x}$

Solution:

Keystrokes

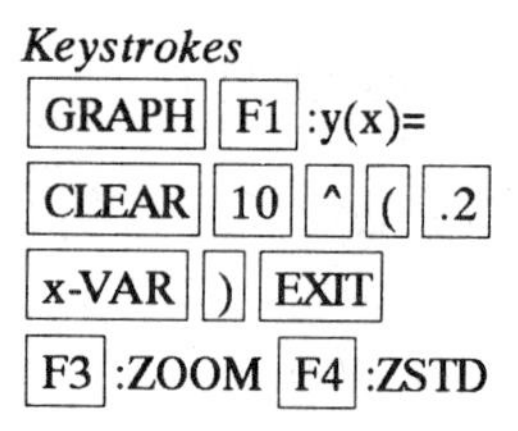

Screen Display

`y1=10^(.2 x)`

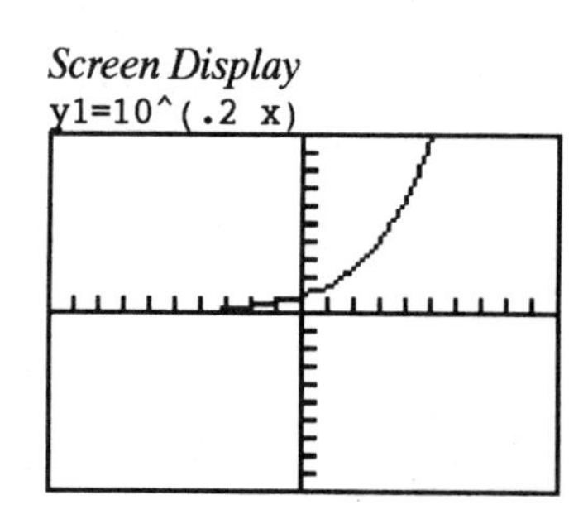

Explanation

Store the function and graph. Note the entire exponent needs to be in parentheses.

Example 2 Graph $y = \frac{e^{x} - e^{-x}}{2}$.

Solution:

Keystrokes

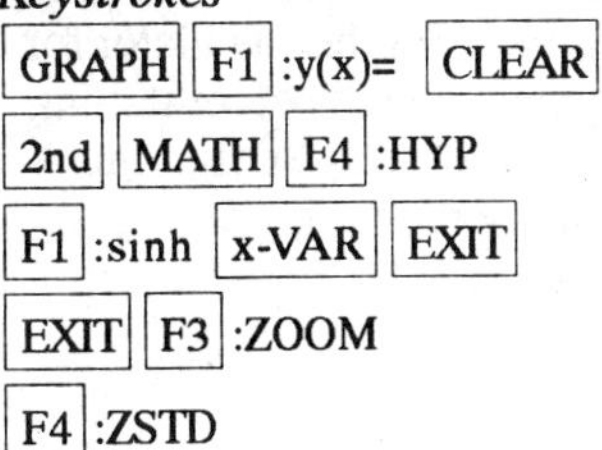

Screen Display

y1=sinh x

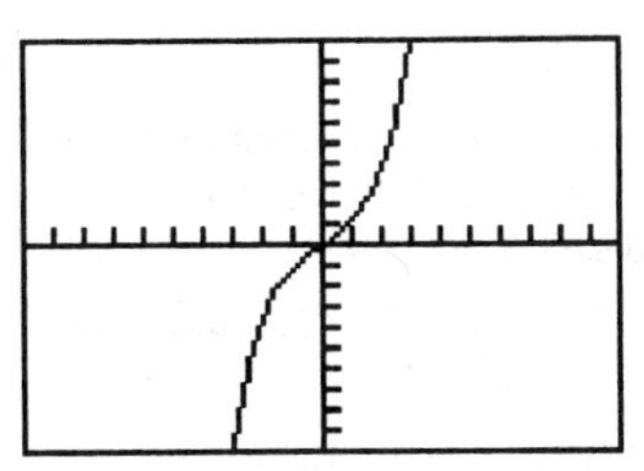

Explanation

We observe that this is the hyperbolic sine function. So we can use the built-in function in the calculator.

The function could also have been graphed by storing:

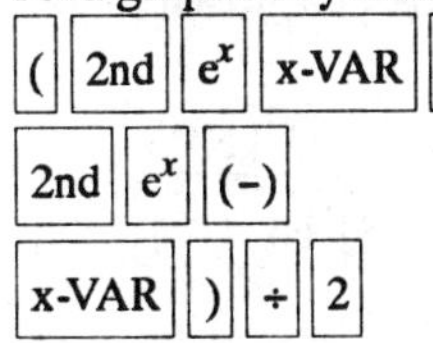

as a function and graphing.

C-18 Angles and Trigonometric Functions

Example 1 Evaluate $f(x) = \sin x$ and $g(x) = \tan^{-1} x$ at $x = \frac{5\pi}{8}$.

Solution:

Keystrokes

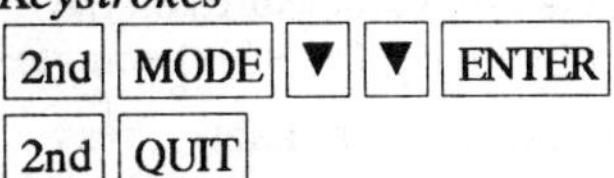

Screen Display

```
Normal Sci Eng
Float 012345678901
Radian Degree
RectC PolarC
Func Pol Param DifEq
Dec Bin Oct Hex
RectV CylV
dxDer1 dxNDer
```

Explanation

The angle measure is given in radians. Set the calculator for radian measure before starting calculations. Return to the Home screen using 2nd QUIT.

Keystrokes	Screen Display	Explanation
5 2nd π ÷ 8 STO▸ ALPHA x-VAR ENTER	5π/8→x 1.96349540849	Store $\frac{5\pi}{8}$ as x.
SIN x-VAR ENTER	sin x .923879532511	Enter f(x) and evaluate.
2nd TAN⁻¹ x-VAR ENTER	$\tan^{-1}$ x 1.09973974852	Enter g(x) and evaluate.

Example 2 Evaluate f(x) = csc x at x = 32° 5' 45".

Solution:

Keystrokes

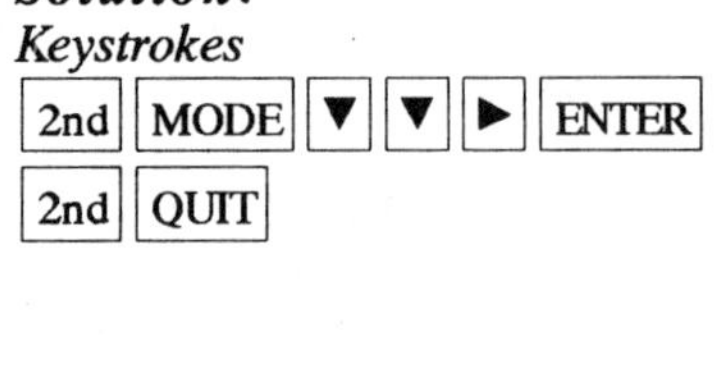

Screen Display

```
Normal Sci Eng
Float 012345678901
Rad Deg
RectC PolarC
Func Pol Param DifEq
Dec Bin Oct Hex
RectV CylV
dxDer1 dxNDer
```

Explanation

The angle measure is given in degrees. Set the calculator for degree measure before starting calculations. Return to the Home screen using 2nd QUIT.

Get ANGLE mode from the MATH menu.

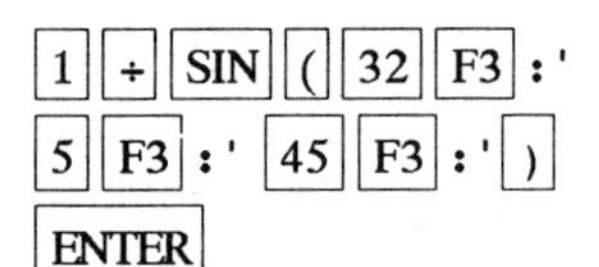


```
1/sin (32'5'45')
          1.88204482194
```

Use 1/sin x as csc x.

Degrees, minutes and seconds can be entered directly using the ' from the MATH menu.

Example 3 Graph f(x) = 1.5 sin 2x.

Solution:

Keystrokes

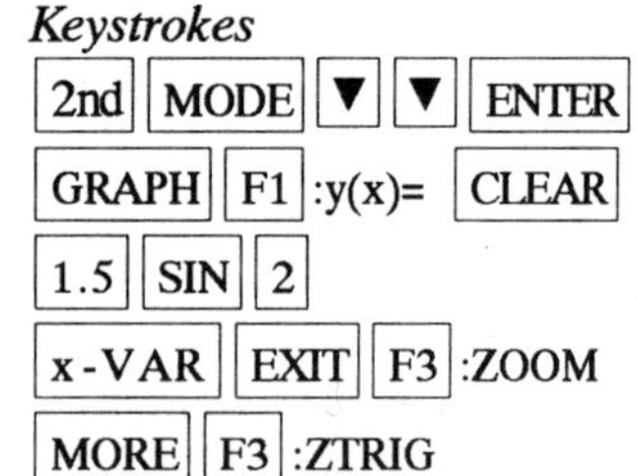

Screen Display

y1=1.5sin 2x

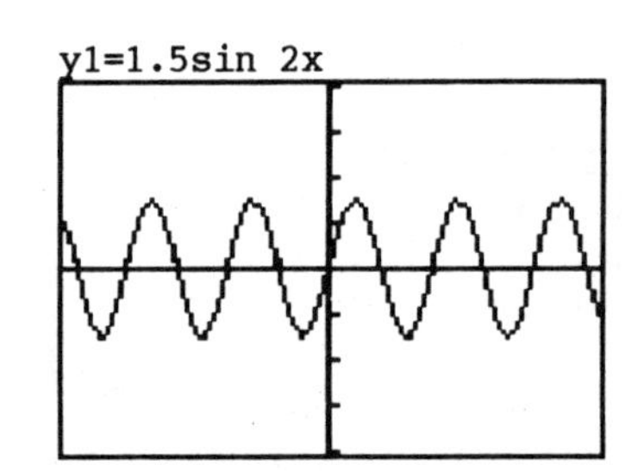

Explanation

Set MODE to radian measure. Store f(x) as y1. Use the trigonometric option on the ZOOM menu to get tick marks set at radian measures on the horizontal axis since the angle measure is in radians. Press F2 :RANGE to see the RANGE is [-8.24..., 8.24...]1.57... by [-4,4]1 on the calculator.

Example 4 Graph g(x) = 3tan^{-1}(.2x).

Solution:

Keystrokes

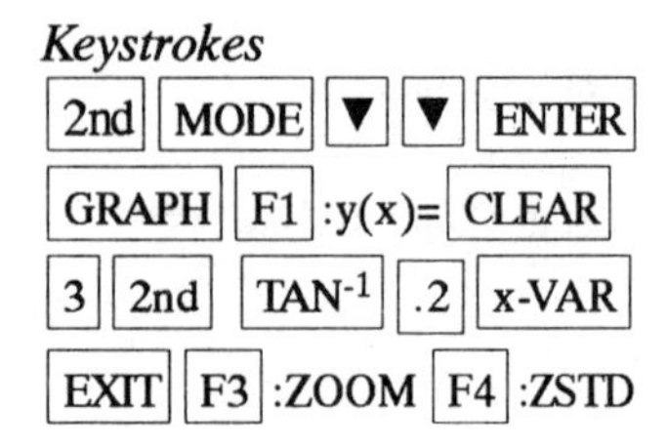

Screen Display

y1=3tan^{-1}.2x

Explanation

Set MODE to radian measure.
Store g(x) as y1.
Use the standard RANGE setting [-10,10]1 by[-10,10]1.

C-19 Polar Coordinates

Example 1 Change the rectangular coordinates $(-\sqrt{3},5)$ to polar form with $r \geq 0$ and $0 \leq \theta \leq 2\pi$.

Solution:

Keystrokes	*Screen Display*	*Explanation*
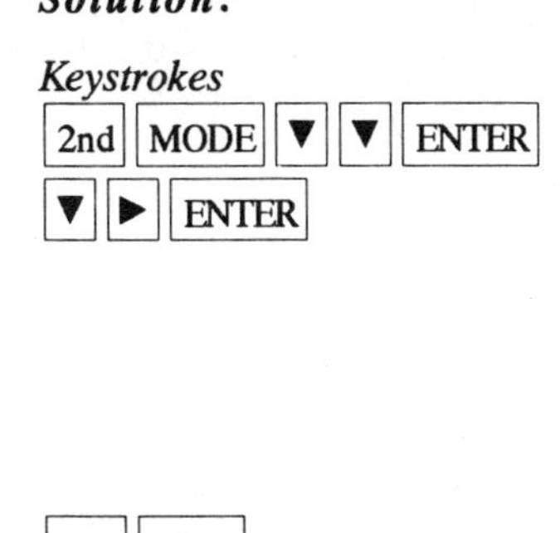	Normal Sci Eng Float 012345678901 Radian Degree RectC PolarC Func Pol Param DifEq Dec Bin Oct Hex RectV CylV dxDer1 dxNDer	Set the mode to Radian angle measure and to PolarC. Now when data is entered in rectangular coordinates, the result will be given in polar coordinates.
		Return to the home screen.
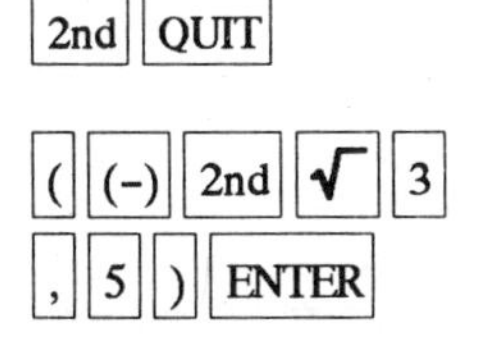	$(^{-}\sqrt{3},5)$ (5.29150262213∠1.904...	Enter the data. The result is in polar coordinates (r,θ). The angle symbol ∠ indicates an angle measure will follow. The calculator will interpret the angle measure to be in radians because we set the mode to radian measure.

Example 2 Change the polar coordinates $(5,\pi/7)$ to rectangular coordinates.

Solution:

Keystrokes	*Screen Display*	*Explanation*
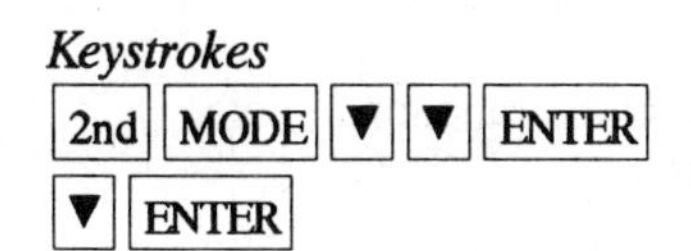	Normal Sci Eng Float 012345678901 Radian Degree RectC PolarC Func Pol Param DifEq Dec Bin Oct Hex RectV CylV dxDer1 dxNDer	Set the mode to Radian angle measure and to RectC. Now when data is entered in polar coordinates, the result will be given in rectangular coordinates.
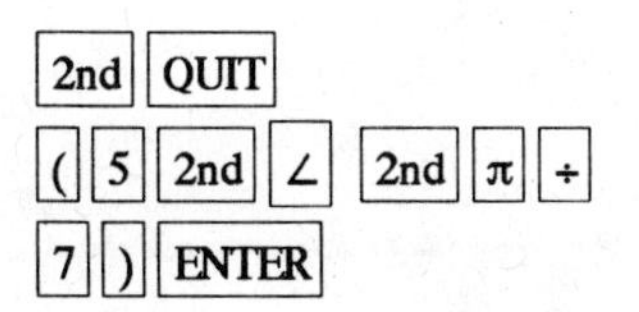	(5∠π/7) (4.50484433951,2.169...	Return to the home screen Enter the polar coordinates. The angle symbol must be used to designate an angle measure is being entered. The result is in rectangular coordinates (x,y)

Example 3 Evaluate $r = 5 - 5\sin\theta$ at $\theta = \frac{\pi}{7}$.

Up to 99 polar equations can be defined and graphed at one time.

Solution:

Keystrokes	*Screen Display*	*Explanation*
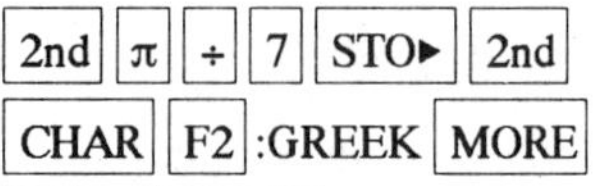	$\pi/7 \rightarrow \theta$ .448798950513	Store $\frac{\pi}{7}$ as θ . θ is on the CHAR menu.
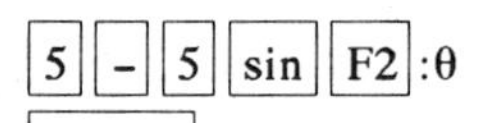	5-5sin θ 2.83058130441	Enter 5-5sin θ and evaluate.

Example 4 Graph $r = 5 - 5\sin\theta$

Solution:

Keystrokes	*Screen Display*	*Explanation*
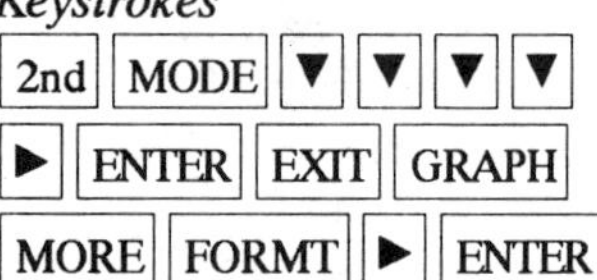		Set the MODE to polar. Set the FORMT to the PolarGC and the rest to default settings (in the leftmost positions). The coordinates shown at the bottom of the screen when using TRACE now will be in polar coordinates.
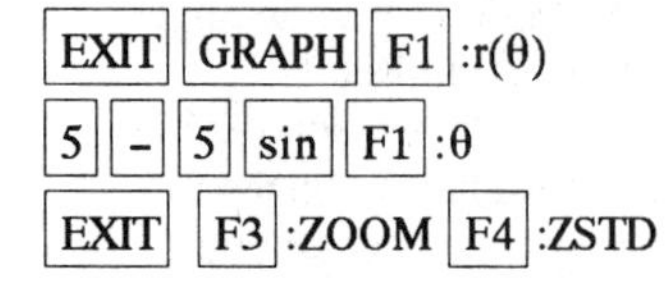	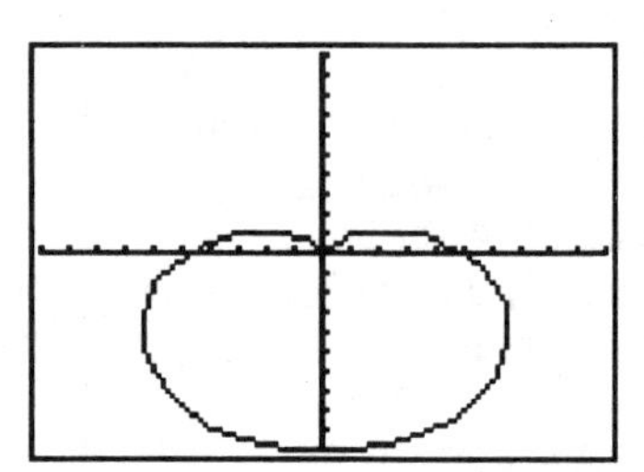	Enter the function in polar form.

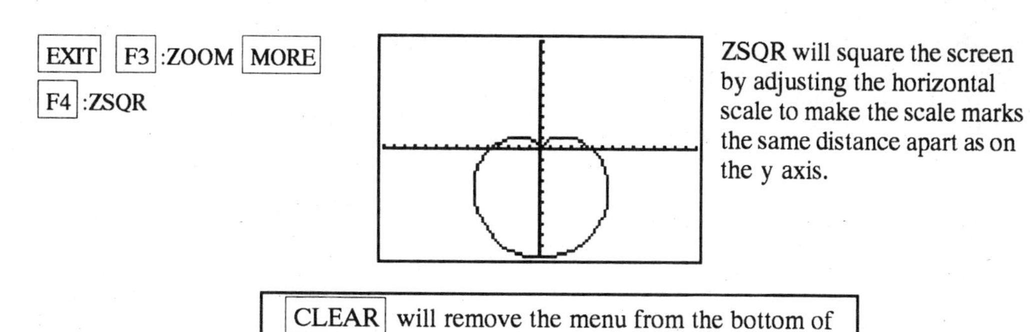

Keystrokes	Screen Display	Explanation
EXIT F3 :ZOOM MORE F4 :ZSQR		ZSQR will square the screen by adjusting the horizontal scale to make the scale marks the same distance apart as on the y axis.

CLEAR will remove the menu from the bottom of the graph screen without removing the graph itself.

C-20 Scientific Notation, Significant Digits, and Fixed Number of Decimal Places

Numbers can be entered into the calculator in scientific notation.

Example 1 Calculate $(-8.513\times10^{-3})(1.58235\times10^{2})$. Enter numbers in scientific notation.

Solution:

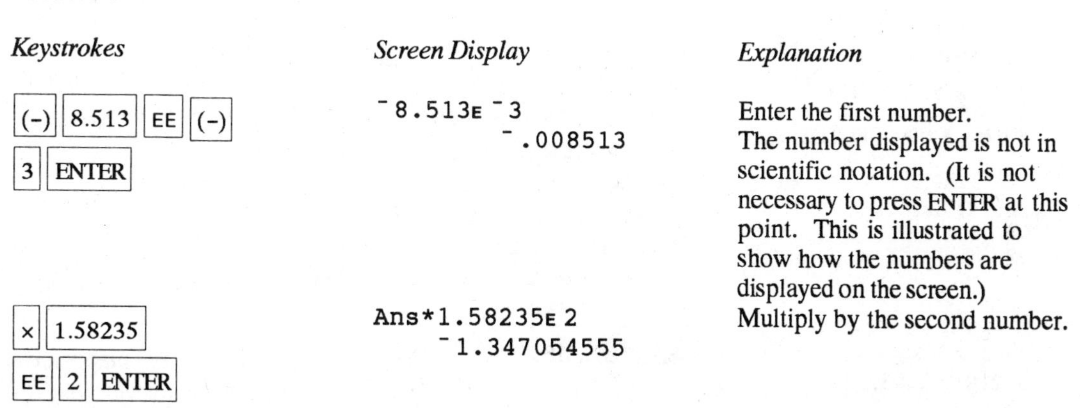

Keystrokes	*Screen Display*	*Explanation*
(-) 8.513 EE (-) 3 ENTER	⁻8.513E⁻3 ⁻.008513	Enter the first number. The number displayed is not in scientific notation. (It is not necessary to press ENTER at this point. This is illustrated to show how the numbers are displayed on the screen.)
× 1.58235 EE 2 ENTER	Ans*1.58235E 2 ⁻1.347054555	Multiply by the second number.

Example 2 Set the scientific notation mode with six significant digits and calculate

$(351.892)(5.32815 \times 10^{-8})$.

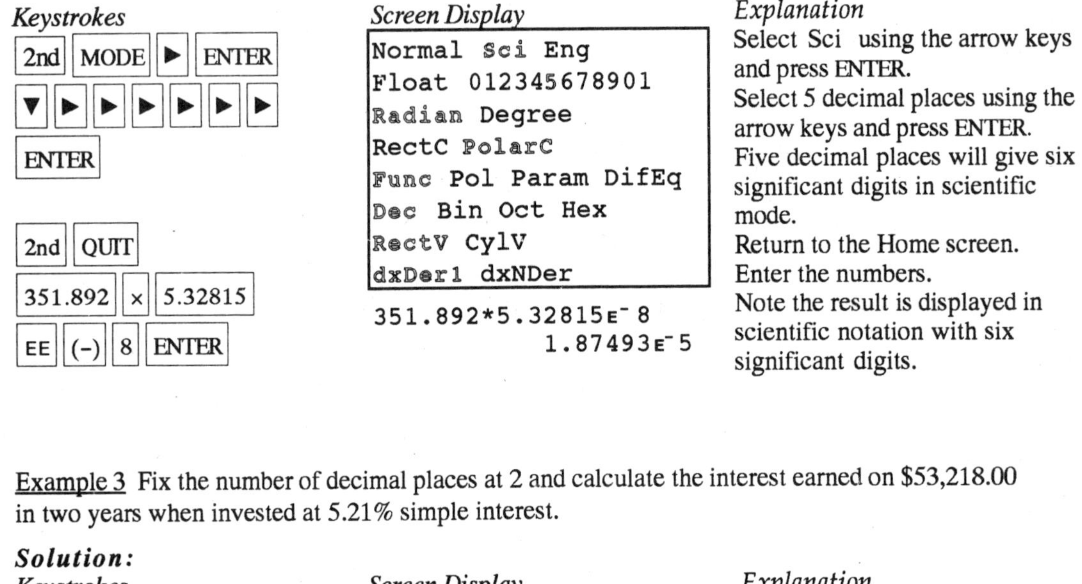

Example 3 Fix the number of decimal places at 2 and calculate the interest earned on \$53,218.00 in two years when invested at 5.21% simple interest.

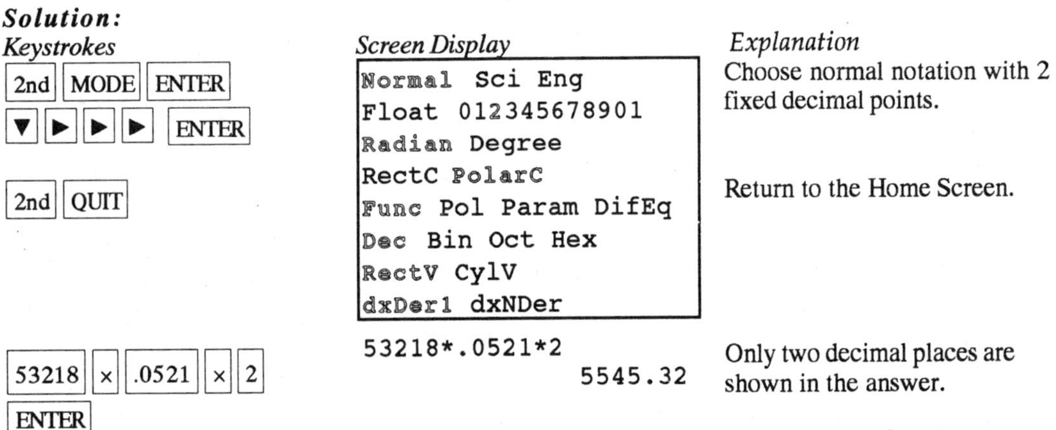

The interest is \$5545.32.

ANSWERS TO SELECTED EXERCISES

Chapter 2

2. (A)

x Record Albums	y Production Cost
2000	26,880
4000	36,080
6000	45,280
8000	54,480

(B)

x Record Albums	y Revenue
2000	16,000
4000	32,000
6000	48,000
8000	64,000

(C) $y = 17680 + 4.60x$
(D) $y = 8x$

(E)

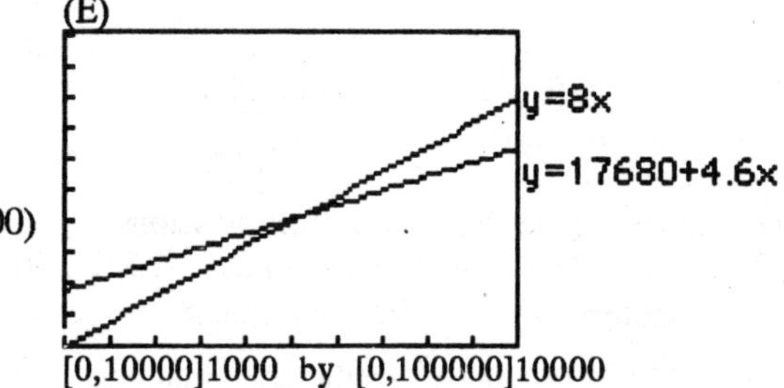

(F) Break-even point is (5200, 41600)
(G) 6875 albums for Revenue = $55000. The profit is $5695.

Chapter 2

1. $a = \pm 2.63$.
2. $a = \pm 1.87$
6.

Textbook problem	Domain	x intercepts	y intercepts	Vertical Asymptotes	Horizontal Asymptotes	Oblique Asymptotes
#11	$x \neq \pm 4$	0 and -1.5	0	$x = \pm 4$	$y = 2/3$	none
#13	all reals	0	0	none	$y = 0$	none
#15	$x \neq -1$, $x \neq 5/3$	0	0	$x = -1$ and $x = 5/3$	none	none

7. #41

(A) (f+g)(-1)=2, (f-g)(-1)=2, (fg)(-1)=0, (f÷g)(-1)=not defined, (f∘g)(-1)=not defined, (g∘f)(-1)=2

(B) (f+g)(2.5)=2, (f-g)(2.5)=-2, (fg)(2.5)=0, (f÷g)(2.5)=0, (f∘g)(2.5)=0, (g∘f)(2.5) is undefined

(C) (f+g)(.035)=2, (f-g)(.035)=-2, (fg)(.035)=0, (f÷g)(.035)=0, (f∘g)(.035)=0, (g∘f)(.035) is undefined

#51

(A) (f+g)(-1)=8.06, (f-g)(-1)=1.74, (fg)(-1)=15.49, (f÷g)(-1)=1.55, (f∘g)(-1)=3.87, (g∘f)(-1)=5.74

(B) (f+g)(2.5)=8.23, (f-g)(2.5)=.43, (fg)(2.5)=16.91, (f÷g)(2.5)=1.11, (f∘g)(2.5)=3.12, (g∘f)(2.5)=5.27

(C) (f+g)(.035)=8, (f-g)(.035)=2.00, (fg)(.035)=15.00, (f÷g)(.035)=1.67, (f∘g)(.035)=4.00, (g∘f)(.035)=5.83

8. Use [-1,1]1 by [0,5]1 to see the graph has a maximum at $x=0$ and a minimum at $x\approx.44$.

Chapter 3

1. The values are the same for both expressions.
3. #25 -7.26, -.77 #27 .68 #29 -1.21, 1.07 #31 -1.69, -.97, 1.17
5. -.3571 multiplicity 1, 1.926 multiplicity 1. [Hint: Graph P(x) and $y=(x+.3571)^r(x-1.926)^q$ for the combinations: $r=1,q=3$ $r=2,q=2$ and $r=3,q=1$. None of the graphs are the same as P(x). Hence multiplicity of both roots is 1 and there are two imaginary roots.

Chapter 4

2. & 3. The functions are reflections about the y axis.

4. The functions have similar shape but 5^x is "steeper" than 3^x. In other words, (5^x is below 3^x for x<0; 5^x is above 3^x for x>0)

5. The functions are reflections about the x axis.

15. #53

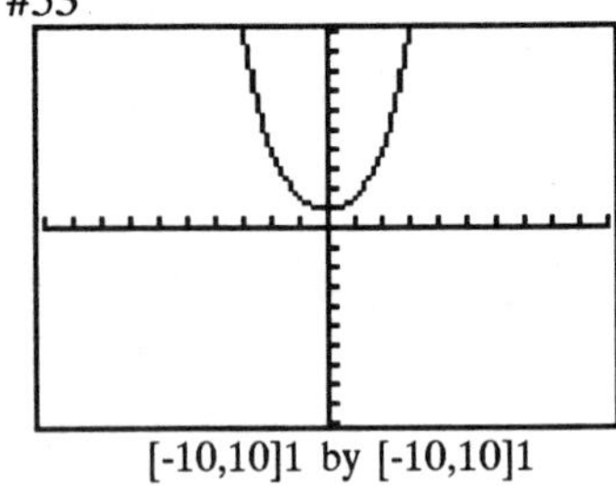

[-10,10]1 by [-10,10]1

#55

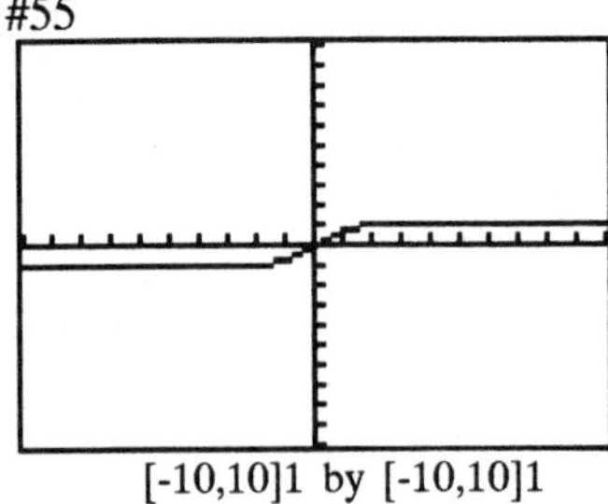

[-10,10]1 by [-10,10]1

Chapter 5

7. The values of cot x are negative with large magnitude for values of x a little less than π.
   The values of cot x are positive with large magnitude for values of x a little greater than π.
8. The values of sec x are positive with large magnitude for values of x a little less than $\pi/2$.
   The values of sec x are negative with large magnitude for values of x a little greater than $\pi/2$.
9. The values of csc x are negative with large magnitude for values of x a little less than 0.
   The values of csc x are positive with large magnitude for values of x a little greater than 0.

Chapter 6

1. Both sides will have the same value.
2. The graphs will be the same for both sides.
3. Both sides will have the same value.
4. The graphs will be the same for both sides.
7. The values will be the same.
8. The graphs will be the same for both sides.
11. Both sides will have the same value.

Chapter 10

1. #39 $-2\frac{2}{3}$ #41 4.02 #43 -12.46
7. (A) $S_{15} = 15.01599$, $S_{50} = 18.12822$. (B) $S_{\infty}=18.18182$ for $r = .89$;
   The errors are: $S_{\infty} - S_{15} = 3.16583$ and $S_{\infty}-S_{50} = .05360$

9. #37 $-9.76751378 \times 10^{10}$ #39 179502.391 #41 18479844.4
   #43 34246040.9

Chapter 12

3. $C(45,10)C(35,5)C(30,8) = 6.06150498 \times 10^{21}$.
4. $C(15,2)C(12,2)(C(9,2))^6 = 1.508510159 \times 10^{13}$
5. $\frac{6}{6^3} = .02\overline{7}$

6. $1/C(8,1) = .125$

- NOTES -

- NOTES -

- NOTES -

- NOTES -

- NOTES -

- NOTES -